The Suppression Lesson: Water vs Fire

Eric J. Neal

Published by Eric J. Neal, 2024.

While every precaution has been taken in the preparation of this book, the publisher assumes no responsibility for errors or omissions, or for damages resulting from the use of the information contained herein.

THE SUPPRESSION LESSON: WATER VS FIRE

First edition. December 20, 2024.

ISBN: 979-8230600206

Written by Eric J. Neal.

Also by Eric J. Neal

From the Streets to a Chief
Igniting Excellence: Challenges of a First-Year Fire Chief
Echoes From the Ashes
The Roadmap to Fire Behavior
The Suppression Lesson: Water vs Fire
From the Shadows: SMU's Journey
Hazardous Materials Unveiled
Rescue Operations Strategies for Saving Lives
Firefighter Survival: Tactics for Staying Alive

To the brave men and women of the fire service,Your unwavering courage, relentless dedication, and selfless acts define the essence of heroism. Each call you answer, every flame you face, and all the lives you protect are a testament to your commitment to a cause greater than yourselves.

This book is dedicated to you—the firefighters who stand ready in the face of danger, embodying resilience, compassion, and an unyielding spirit. May it serve as a tool to enhance your knowledge, support your mission, and honor your incredible sacrifice.

Thank you for being the protectors of our communities and the guardians of hope.

Chapter 1: Fire Suppression Techniques

Suppressing fires is a complex endeavor that requires more than just bravery and a hose. It encompasses a deep understanding of fire behavior, employing precise methods to tackle different types of fires while safeguarding the lives of both firefighters and civilians. The art of battling blazes lies in recognizing how fires start, spread, and can be effectively extinguished—each scenario presenting its unique challenges that demand tailored responses. Grasping these elements transforms what might appear as chaos into a comprehensible pattern, enabling strategic thinking and efficient actions on the ground. When armed with this knowledge, firefighters can approach incidents not just reactively but with foresight and control, elevating their effectiveness in preserving both life and property.

Within this chapter, we delve into the intricacies of fire suppression, exploring vital concepts such as the fire triangle's role in sustaining combustion and how disrupting its components—fuel, heat, and oxygen—can halt the flames. Readers will journey through an examination of the various classes of fires and why each demands specific tactics, ensuring safety and efficacy in response efforts. Beyond theoretical insights, practical strategies will be unveiled, shedding light on cooling techniques, smothering methods, and fuel removal strategies, each meticulously designed to adapt to the unpredictable nature of fire. Moreover, safety considerations are underscored throughout, emphasizing the delicate balance between action and precaution that defines successful firefighting operations. By the end of this exploration, readers should find themselves equipped with a clearer understanding of what it means to suppress a fire safely and successfully, gaining insights that translate beyond theory into real-world applications on the frontline.

Understanding Fire Behavior

UNDERSTANDING FIRE behavior is essential for effective suppression, providing firefighters with the knowledge needed to tackle various situations strategically. At the heart of this understanding is the combustion process, a chemical reaction that requires three key elements: fuel, heat, and oxygen. Consider it as a fire triangle where each side supports the other two; disrupting any one element can extinguish the fire.

First, let's explore how removing components of this triangle works in real-life scenarios. For instance, when a fire runs out of fuel, it simply can't continue to burn. This principle is why clearing brush away from homes in wildfire-prone areas is an effective measure. Similarly, heat removal, often achieved by applying water to absorb caloric energy, prevents the temperature from reaching levels necessary for ignition. Removing oxygen, another tactic, involves using smothering agents like foam, which can effectively starve a fire of the air it needs to sustain itself.

However, it's not just about knowing what fires need; understanding the different classes of fire also plays a critical role. Firefighters deal primarily with Class A fires—those involving ordinary combustibles such as wood, paper, or cloth, and Class B fires, which involve flammable liquids and gases. Each class requires specific responses due to differing materials and combustion properties. For example, while water might be effective against a pile of burning magazines (Class A), it's not suited for a gas leak explosion (Class B), where specialized foams are preferential. This tailored response ensures maximum safety and efficiency in suppression efforts.

Moving on, consider how heat travels and spreads fire. Heat transfer occurs through conduction, convection, and radiation, all of which dictate how a fire behaves and escalates. In conduction, heat travels through materials. Think of metal beams in a building; they can carry heat from one area to another, potentially igniting new fires distant from the original source. Convection involves the movement of hot gases and smoke within a fluid medium like air or liquid, akin to how warm air rises from a campfire, spreading heat upwards and sometimes horizontally. Radiation, meanwhile, transmits heat via electromagnetic waves, allowing fire to ignite items without direct contact. Imagine the way intense heat can melt window frames without flames touching them directly.

Recognizing these mechanisms allows firefighters to predict fire paths and effectively strategize containment efforts. By anticipating where a fire might spread next, personnel can create barriers, deploy cooling measures, or position teams safely out of harm's way while preparing to combat the advancing blaze. Understanding fire behavior extends beyond immediate flames to include environmental factors like wind patterns, ambient temperatures, and moisture levels. These elements influence how quickly a fire grows or changes direction, thus shaping suppression strategies. In wildland fire scenarios, wind can drive flames at alarming speeds, while high temperatures and low humidity exacerbate conditions, making suppression more challenging.

Successful fire suppression hinges on recognizing these behavioral nuances and adapting accordingly. It means gathering as much data as possible about the incident at hand, including visual assessments and technological readings, to refine approaches during operations. This information supports real-time decision-making, ensuring that actions taken are both informed and decisive.

Moreover, a nuanced understanding of fire interactions aids in developing comprehensive suppression plans. Tactics must be flexible enough to adjust for unexpected changes while steadfastly focused on overarching safety goals. By piecing together insights into combustion processes, fire classes, heat transfer, and environmental impacts, firefighters can mount a coordinated defense tailored to suppress specific threats.

Effective firefighting becomes a dance between observation and action, requiring discipline and adaptability. Knowledge of fire behavior empowers firefighters to engage confidently, protecting not only property but also human lives. As training evolves and technology advances, the importance of grasping these foundational principles remains unchanged, underpinning every successful firefighting operation with a bedrock of understanding.

Methods of Fire Suppression

IN FIRE SUPPRESSION, understanding the primary methods used is critical for effective firefighting. These techniques vary depending on the nature and classification of the fire but generally focus on disrupting the essential components that sustain a fire: heat, oxygen, and fuel.

Cooling is perhaps the most straightforward suppression method. It involves using water to absorb heat from the fire, effectively lowering its temperature below ignition points. This method is particularly efficient when dealing with Class A fires, which involve ordinary combustibles like wood, paper, and textiles. Water's high heat capacity makes it an ideal agent for cooling since it can absorb significant amounts of heat quickly. When applied, water not only cools the flames but also helps minimize the production of flammable vapors, facilitating a safer environment for firefighters. However, it's important to remember that water should never be used in scenarios involving electricity or flammable liquids, as it can exacerbate the situation (*What Are the 3 Methods for Extinguishing a Fire?*, 2023).

Smothering is another vital technique in fire suppression. By cutting off the fire's oxygen supply, the combustion process is effectively halted. This approach is commonly used for Class B fires, which are characterized by flammable liquids such as gasoline, oils, and greases, and Class K fires found in kitchens involving cooking oils. Foam extinguishers, for example, create a barrier between the fuel and the air, preventing oxygen from reaching the fire. The foam blankets the burning material, suppressing vaporization, and subsequently cutting off the oxygen supply, thereby extinguishing the fire. Smothering can also involve other innovative approaches like using a fire blanket or encasing the area in a controlled environment where oxygen levels are drastically reduced (Foster, 2021).

Fuel removal is a suppression strategy that focuses on isolating or eliminating the fire's fuel source. Particularly crucial in industrial settings and wildland fires, this method requires strategic planning and execution. In industrial environments, shutting off valves to stop the flow of flammable liquids or gases can dramatically reduce the available fuel, thus containing the fire. In forested areas, creating firebreaks or controlled backburning can remove vegetation that serves as potential fuel, making it harder for the fire to spread. Such tactical fuel removal not only forms a physical barrier but also provides a psychological advantage to firefighters, as they can control the fire's progression through careful resource allocation and timely intervention (*What Are the 3 Methods for Extinguishing a Fire?*, 2023).

Chemical flame inhibition is another effective method, particularly suited for combating Class B and C fires, which involve electrical hazards. This technique employs chemical agents such as dry powders or gases to interfere with the chemical reactions occurring during combustion. Dry chemical extinguishers, for instance, release substances like monoammonium phosphate or sodium bicarbonate that disrupt the flame's chemical processes, effectively suffocating the fire. These agents act rapidly to neutralize the free radicals in the combustion process, halting the fire's ability to sustain itself. Chemical agents are invaluable in situations where water and foam are either ineffective or unsafe to use, such as in electrical fires, where conductivity poses a risk of electrocution (Foster, 2021).

Safety in Fire Suppression

IN FIREFIGHTING, SAFETY measures are paramount to successful fire suppression. One of the first steps in ensuring safety during fire suppression is conducting a thorough scene assessment. Firefighters must identify potential hazards that could threaten their safety or impair their operations. Structural instability is one such hazard; buildings affected by fire can become compromised, with weakened beams and floors posing significant threats. Identifying hazardous materials is equally crucial as fires in these environments can lead to toxic exposures or dangerous reactions.

Proper Personal Protective Equipment (PPE) forms the frontline defense against these risks. Turnout gear and Self-Contained Breathing Apparatus (SCBA) are indispensable in shielding firefighters from burn injuries and inhalation of toxic smoke. According to IFTSA Firefighter Principles & Tactics, today's bunker gear provides limited protection in extreme conditions like flashovers, emphasizing the importance of understanding its limits and not becoming over-reliant on the gear for survival (Staff, 2014). PPE should be worn correctly and checked regularly, as even slight faults can expose firefighters to severe harm.

Continuous monitoring of fire conditions is essential to maintaining safety. Firefighters need to remain vigilant about changing conditions that may signal imminent dangers such as flashovers, which occur when a room's contents reach ignition point simultaneously due to high temperatures. Recognizing signs of a potential backdraft—an explosion caused when oxygen is suddenly reintroduced to an oxygen-deprived fire environment—is also critical. Using thermal imaging cameras and honing observational skills to read smoke color, volume, and movement can provide early warnings, allowing crews to adjust tactics accordingly.

Implementing safe procedures involves more than just equipment; it encompasses strategies to mitigate risk while still performing necessary actions effectively. Standard operating procedures should be adhered to rigorously. For instance, controlling airflow to a fire can significantly alter how the fire behaves. This includes strategic ventilation, where openings are created deliberately to control fire spread rather than unintentionally feeding it with additional oxygen. Additionally, ensuring a charged hoseline is ready before beginning tactical ventilation helps prepare for sudden changes in fire conditions (Staff, 2008).

Establishing and maintaining communication among crew members cannot be overstressed. Clear communication ensures everyone is aware of the current situation, planned actions, and potential escape routes if conditions deteriorate. The phrase "If you see something, say something" embodies this approach—any team member noticing signs of danger needs to alert others immediately. This proactive communication culture allows teams to regroup quickly and adapt their strategies, preventing tragedies born out of miscommunication or neglect.

Escape plans and continuous situational awareness are pivotal for firefighter safety. Alternate exits should be identified as soon as possible upon arrival at the scene, providing options if primary routes are blocked or unsafe. Areas of refuge within the structure should also be noted, offering temporary safety until conditions stabilize or further assistance arrives. Effective training predisposes firefighters to operate under these guidelines instinctively, enhancing their ability to react promptly and decisively.

Bringing It All Together

IN THIS CHAPTER, WE'VE delved into the essentials of understanding fire behavior to enhance fire suppression efforts. By dissecting the combustion process, we grasp the crucial elements—fuel, heat, and oxygen—that sustain a fire. Recognizing how these components interact allows firefighters to strategically disrupt the fire triangle, whether it's through removing fuel, reducing heat, or cutting off oxygen. Additionally, identifying different classes of fires and their unique properties ensures that appropriate suppression methods are selected for each scenario, maximizing safety and effectiveness. Understanding the ways heat transfers—through conduction, convection, and radiation—is also key in predicting fire movement and strategizing containment efforts.

Safety is interwoven throughout these strategies, ensuring the well-being of those on the frontline. This chapter underscores the importance of thorough scene assessments, the use of Personal Protective Equipment (PPE), and continuous monitoring of conditions to preemptively address potential hazards like structural instability or toxic exposures. By following safe procedures, maintaining clear communication, and preparing escape routes, firefighters can adapt to dynamic environments with confidence. The insights offered here cement a foundation upon which successful firefighting operations stand, blending knowledge with action to protect lives and property effectively.

Chapter 1 Questions and Answers

What are the three essential components of the fire triangle?

Fuel, heat, and oxygen.

What happens when one component of the fire triangle is removed?

The fire is extinguished because combustion can no longer be sustained.

Why is understanding the fire triangle important for suppression tactics?

It helps firefighters choose the most effective method—cooling, smothering, or fuel removal—to disrupt combustion.

What class of fire involves ordinary combustibles like wood or paper?

Class A fires.

Why is water ineffective or dangerous on Class B fires?

Because Class B fires involve flammable liquids and gases, and water can spread the flammable substance or react violently.

How does heat transfer via conduction affect fire behavior?

It allows heat to travel through solid materials like metal, potentially igniting distant areas.

What role does convection play in the spread of fire?

It moves hot gases and smoke, allowing heat to rise and spread flames vertically and horizontally.

How does radiation contribute to fire spread?

It transmits heat through electromagnetic waves, igniting materials without direct contact.

Name one environmental factor that can influence fire behavior.

Wind patterns.

Why is anticipating fire spread crucial for suppression efforts?

It enables firefighters to position resources and create barriers proactively for safety and containment.

What is the primary method used to cool fires and reduce heat?

Applying water.

Why is water effective for Class A fires?

Because it absorbs heat quickly and reduces the fire's temperature below ignition point.

What suppression method involves cutting off a fire's oxygen supply?

Smothering.

How do foam extinguishers work?

They form a barrier that prevents oxygen from reaching the fire and suppress vapor release.

What is fuel removal, and when is it commonly used?

It involves eliminating the fire's fuel source and is commonly used in wildland or industrial fires.

What is chemical flame inhibition?

A method that interrupts the chemical reaction in combustion using agents like dry powders or gases.

Why are chemical agents preferred for Class C (electrical) fires?

Because they do not conduct electricity and can safely extinguish the fire.

Why is PPE vital in fire suppression operations?

It protects firefighters from burns, toxic smoke, and other hazards on the scene.

What tools or observations help identify potential flashovers or backdrafts?

Thermal imaging cameras and smoke analysis (color, volume, movement).

How does communication contribute to firefighter safety?

It ensures all team members are informed of evolving conditions, tactical plans, and escape routes.

Chapter 2: Nozzle Selection and Applications

Choosing the right nozzle in firefighting can significantly influence the effectiveness of fire suppression efforts. It's not just about having a nozzle but understanding why each type of nozzle is crafted the way it is and how it can best be applied to various firefighting scenarios. Out on the field, where seconds matter and tactics can shift at a moment's notice, knowing which tool to wield is critical. The nozzles in your firefighting arsenal are designed with specific attributes tailored for different challenges—whether it's maximizing reach, providing flexibility, or ensuring consistent performance. This decision isn't simply technical; it's strategic, as the selection can dictate the difference between controlling a blaze swiftly or facing prolonged engagement.

In this chapter, we delve into the distinct types of nozzles available, examining their capabilities and applications in depth. You will uncover the power behind Smooth Bore Nozzles, appreciated for their durability and solid streams that penetrate intense infernos effectively. Explore how Combination or Fog Nozzles provide crucial versatility, allowing firefighters to adapt stream patterns to suit varied situations, including producing cooling fogs in enclosed spaces. The innovative Automatic Nozzles will also be discussed, showcasing their self-adjusting abilities that maintain steady pressure despite fluctuating water supply conditions. Additionally, the nuances surrounding nozzle pressure and flow rates, along with their impact on firefighting strategies, will be explored, providing insights into aligning these factors with operational needs. This chapter serves as an essential guide to mastering effective nozzle use, enhancing both tactical approaches and safety outcomes in firefighting operations.

Types of Nozzles

SELECTING THE RIGHT nozzle in firefighting is pivotal for effective fire suppression. Among the various types of nozzles available, each has specific attributes that make it suitable for different scenarios. In this discussion, we will delve into three significant nozzle types used extensively in firefighting: Smooth Bore Nozzles, Combination/Fog Nozzles, and Automatic Nozzles.

Firstly, let's explore Smooth Bore Nozzles, a favorite among many firefighters for their ability to deliver powerful, solid streams of water. These nozzles are known for their high-impact penetration capacities, making them indispensable when confronting intense fires. They work best in situations that demand substantial flow rates and long-distance reach. The smooth bore design minimizes friction loss, which means more water reaches the fire with less pressure required from the pump. This can be particularly beneficial when dealing with large blazes where water supply might be an issue. The straightforward design of Smooth Bore Nozzles—lacking moving parts—also makes them durable and easy to maintain in tough environments. While they excel in delivering concentrated water streams, these nozzles do come with some limitations, as they primarily operate in one stream pattern and do not break up the water into smaller droplets to effectively absorb heat from the surrounding environment (Staff, 2012).

On the other hand, Combination or Fog Nozzles bring versatility to firefighting efforts with their adjustable stream patterns. Unlike Smooth Bore Nozzles, Fog Nozzles can produce a wide spectrum of spray patterns, from a tight straight stream to a broad fog. This adaptability is crucial for executing versatile fire attacks, especially in scenarios requiring both direct firefighting and cooling hot gases. For instance, in enclosed spaces where heat buildup and smoke present significant hazards, Fog Nozzles can create a protective water curtain by dispersing fine droplets that rapidly absorb heat and reduce the risk of flashover—a critical tactic in combating indoor fires. Moreover, the capability to adjust the nozzle's pattern makes it ideal for both offensive and defensive firefighting strategies, allowing firefighters to tailor their approach based on the intensity of the flames and the layout of the building. However, operating a Fog Nozzle typically requires higher pressure compared to a Smooth Bore Nozzle, which can lead to challenges if the water source is inconsistent (Bettinazzi, 2023).

Automatic Nozzles introduce another dimension altogether with their self-adjusting features designed to maintain consistent pressure while allowing variable flow rates. This type of nozzle is particularly advantageous in circumstances where the water supply fluctuates, ensuring that firefighters maintain effective control over the water output without needing constant manual adjustments. The automatic mechanism within these nozzles compensates for changes in pressure, providing reliability during dynamic firefighting operations. This feature can prove critical in unpredictable fire scenarios where water availability may change suddenly, such as in rural areas or during prolonged engagements. Despite their sophisticated technology, automatic nozzles must be understood thoroughly by operators to prevent malfunctions. Incorrect use can lead to improper flow rates, compromising their effectiveness in dire situations (Staff, 2012).

Each of these nozzle types comes with a unique set of characteristics tailored to specific firefighting needs. Smooth Bore Nozzles offer simplicity and efficiency in delivering strong streams, excelling in open areas where reach is paramount. Their ability to transport large volumes of water with minimal resistance makes them reliable in reducing high-temperature zones. Fog Nozzles, with their expansive coverage options, cater to varying tactical requirements, providing essential protection and cooling capabilities in confined spaces. Their utility extends beyond structural fires, finding use in wildland settings where preventing flare-ups and controlling perimeter fires are necessary. Finally, Automatic Nozzles represent technological advancement, balancing between handling diverse pressures and sustaining consistent and safe water application, thus reinforcing operational stability under extreme conditions.

Advantages and Disadvantages of Nozzle Types

IN THE WORLD OF FIREFIGHTING, nozzle selection is crucial. Each type of nozzle offers unique advantages and disadvantages based on its design and intended use. Understanding these pros and cons can greatly influence the effectiveness of fire suppression efforts and help firefighters select the best equipment for specific scenarios.

Smooth Bore Nozzles are often celebrated in the firefighting community for their simplicity and efficiency. They produce a solid stream of water, offering greater reach compared to other types of nozzles. This is particularly useful when fighting fires from a distance or needing to target specific areas with high impact. For example, in large open spaces or when battling flames that originate from elevated positions, the reach provided by Smooth Bore Nozzles can be indispensable. Additionally, these nozzles require lower nozzle pressure, which translates into less strain on pumping equipment and reduces the physical burden on firefighters due to decreased nozzle reaction (Bettinazzi, 2023). However, the drawback lies in their limited stream pattern options. Unlike other nozzles, they offer little flexibility in adjusting the stream to suit varying situations. Their inability to easily switch from a direct stream to a spray pattern can be a limitation in scenarios where such adaptability might be beneficial.

Conversely, Combination or Fog Nozzles provide versatility in operations. These nozzles can adjust the stream pattern by rotating the bumper, allowing firefighters to choose between a narrow fog, wide fog, or straight stream (Staff, 2012). This versatility proves advantageous in many firefighting scenarios, such as gas cooling, where breaking the water into fine droplets increases the surface area, enhancing heat absorption. The ability to switch to a fog pattern also aids in firefighter safety by providing a protective water curtain.

However, this versatility comes with its own set of challenges. Combination/Fog Nozzles typically require higher nozzle pressures to function optimally, which could mean more exertion on pumps and personnel. Furthermore, while their reach is decent, it's generally less than that offered by Smooth Bore Nozzles, potentially limiting their effectiveness in long-distance firefighting operations (Bettinazzi, 2023).

Automatic Nozzles bring another dimension to nozzle technology with their adaptive design that automatically adjusts to maintain consistent pressure while allowing variable flow rates. This feature is particularly useful when dealing with fluctuating water supply conditions. They simplify the operation by ensuring the desired pressure is maintained without manual adjustments (Staff, 2012). Despite these advantages, Automatic Nozzles are not without faults. Their complex designs mean there is a need for regular maintenance to ensure they function correctly. Improper handling or infrequent upkeep can lead to improper flow rates, undermining their intended benefits. Additionally, they are generally more expensive than other types, both in initial purchase and ongoing upkeep.

For fire services looking to optimize their resources, understanding how each nozzle performs under various conditions is key. When considering Gallon Per Minute (GPM) delivery, it is essential to ensure that the chosen nozzle aligns with the fire's demands and the available water supply. For instance, a Smooth Bore Nozzle might excel in scenarios where maximum reach and penetration are critical. In contrast, a Fog Nozzle would be better suited for environments requiring rapid cooling and versatility. Matching nozzles with specific fire conditions enhances operational efficiency and safety. For example, in confined spaces where visibility and protection are crucial, the ability to quickly adjust the stream to a fog pattern can shield firefighters from intense heat.

In practical application, selecting the right nozzle requires an understanding of both the fire scenario and the equipment. Firefighters must be trained not only in the mechanical operation of these tools but also in strategic deployment. Testing and familiarization drills can help crews develop proficiency in switching between nozzle types efficiently, depending on situational requirements. A comprehensive evaluation of each nozzle type's strengths and weaknesses can inform decision-making processes during emergency response, ultimately contributing to more effective fire suppression and improved safety outcomes.

Nozzle Pressure and Flow

IN THE WORLD OF FIREFIGHTING, understanding how to manipulate water efficiently can be just as crucial as having a strong fire response team. One key component in this process is the proper selection and utilization of nozzles, with particular emphasis on nozzle pressure and flow rate. These elements are essential for effective fire suppression and maximizing the impact of every water droplet delivered to the fire ground.

The relationship between nozzle pressure and achieving optimal stream patterns cannot be understated. Nozzle pressure directly influences the shape and reach of the water stream, making it a critical factor for firefighters to consider when selecting equipment. Different types of nozzles require varied levels of pressure to perform effectively. For instance, smooth bore nozzles, known for their simple design and solid stream production, operate best at pressures around 50 psi (Bettinazzi, 2023). On the other hand, fixed-gallonage combination fog nozzles have predefined operating pressures that typically range from 50 to 100 psi, depending on the model (Bettinazzi, 2023).

Choosing the correct pressure ensures that firefighters can produce the desired stream pattern, whether as a straight stream for long-distance penetration or a wide spray for greater coverage.

Maintaining adequate pressure prevents issues such as hose kinking, which occurs when there is insufficient pressure leading to a collapse in hose structure and poor maneuverability (Bettinazzi, 2023).

Another crucial aspect is the alignment of Gallon Per Minute (GPM) delivery with fire conditions and available water supply. The GPM rate determines the volume of water dispatched onto the fire, which must align with both the severity of the fire and the capabilities of the local water supply. In situations where fire intensity is high, a higher GPM is often necessary to absorb more heat and create a more significant cooling effect. For example, increasing the nozzle pressure to 60 psi can raise the GPM, delivering more water rapidly, which is ideal for advanced fire scenarios (Bettinazzi, 2023). Conversely, in situations such as overhaul or trash fires, where lower water volume suffices, reducing the pressure to around 40 psi can still be effective while minimizing nozzle reaction and handling difficulties (Bettinazzi, 2023).

It's also vital to match the type of nozzle to the specific fire conditions to maximize efficiency. Smooth bore nozzles are particularly advantageous in scenarios requiring deep penetration and minimal air entrainment. Such characteristics make them suitable for large open spaces or single-point ventilation operations within compartmentalized areas. Their ability to maintain a consistent stream without risk of accidental pattern changes makes them a go-to choice for operations where precision is paramount (Bettinazzi, 2023).

On the other hand, fog nozzles offer versatility, producing adjustable patterns ranging from a fine mist to a straight stream. They are effective in confined spaces or settings where vapor suppression is critical, thanks to their ability to increase surface area exposure by dispersing water into tiny droplets. This characteristic enhances heat absorption, providing an added advantage in controlling the fire (KNOW YOUR NOZZLES, n.d.). Fixed-gallonage fog nozzles, in particular, offer predictability with a constant flow rate, allowing efficient water application even under varying pressure conditions (KNOW YOUR NOZZLES, n.d.).

Understanding these nuances in nozzle operation allows firefighters to tailor their approach based on situational needs, ultimately driving effective fire suppression tactics. It underscores the importance of training and familiarity with nozzle options, prompting firefighters to remain adaptable and prepared for diverse challenges.

By mastering the intricacies of nozzle pressure and flow, firefighters enhance their capability to control and extinguish fires effectively. The ability to switch seamlessly between different nozzle types and adjust for varying pressures enables more precise fire attack strategies and resource management. Therefore, the knowledge and application of these principles not only optimize performance but also contribute significantly to firefighter safety and success in their mission.

Final Thoughts

THROUGHOUT THIS CHAPTER, we have explored the critical role that selecting and using the right nozzles plays in effective fire suppression. Smooth Bore Nozzles stand out for their ability to deliver strong, penetrating streams, ideal for situations where reach is necessary. Their straightforward design makes them both reliable and easy to maintain, although they lack flexibility in stream patterns. Combination or Fog Nozzles offer the versatility needed to adapt to various fire scenarios, from producing fine mists for cooling to forming protective water curtains in enclosed spaces. Yet, their higher pressure requirements can pose challenges, especially when water resources are inconsistent. Automatic Nozzles, with their self-adjusting features, simplify operations by maintaining consistent pressure regardless of water supply fluctuations. However, they demand thorough understanding and maintenance to ensure proper function.

Understanding nozzle pressure and flow is equally crucial, as these elements dictate the effectiveness of each firefighting operation. The correct balance between pressure and flow ensures that firefighters can deliver the appropriate volume and pattern of water to tackle specific fire conditions. Each type of nozzle has its own pressure requirements and capabilities, making it essential for firefighters to be well-trained and familiar with how to adjust these settings accordingly. By mastering these tools and their nuances, firefighters can develop strategies that not only optimize their efforts on the ground but also enhance safety measures during emergency responses. Ultimately, the knowledge gained from learning about nozzle types, pressures, and flows equips firefighters with the skills necessary to approach any fire scenario with confidence and precision.

Chapter 2 Questions and Answers

What are the three main types of firefighting nozzles discussed in Chapter 2?

Smooth Bore Nozzles, Combination/Fog Nozzles, and Automatic Nozzles.

What type of stream does a Smooth Bore Nozzle produce?

A solid, straight stream of water.

What is the main advantage of a Smooth Bore Nozzle in terms of pressure and reach?

It requires less pressure to achieve a long-distance, high-impact stream.

Why are Fog Nozzles preferred in enclosed space fires?

Because they produce a fine mist that absorbs heat quickly and reduces the risk of flashover.

How do Combination/Fog Nozzles offer versatility on the fireground?

They can switch between straight stream, narrow fog, and wide fog patterns.

What makes Automatic Nozzles unique compared to other types?

They self-adjust to maintain consistent pressure despite fluctuations in water supply.

What is one operational drawback of Automatic Nozzles?

They require thorough training and maintenance to avoid malfunction and ensure correct flow rates.

What stream pattern is typically not available with a Smooth Bore Nozzle?

A fog or spray pattern.

What is a disadvantage of using a Fog Nozzle in long-distance firefighting?

Its reach is generally less than that of a Smooth Bore Nozzle.

What operational benefit does a Smooth Bore Nozzle offer regarding nozzle reaction?

Lower pressure requirements reduce nozzle reaction and strain on the operator.

What is a major downside of Fog Nozzles when it comes to water pressure needs?

They require higher nozzle pressures, which can burden pumping systems.

Why are Automatic Nozzles often more expensive than other types?

Due to their complex design and the need for ongoing maintenance.

How does nozzle pressure affect the water stream's reach and pattern?

Higher pressure increases reach and allows pattern control, while lower pressure can reduce performance and stream integrity.

What is the typical operating pressure for a Smooth Bore Nozzle?

Around 50 psi.

What does GPM stand for, and why is it important?

Gallons Per Minute; it determines the volume of water delivered to the fire.

What happens if nozzle pressure is too low?

Stream reach is compromised, and hose kinking may occur, reducing water delivery.

In what type of fire scenario is a Smooth Bore Nozzle most effective?

Large, open-area fires or fires requiring high-penetration from a distance.

How can a Fog Nozzle enhance firefighter safety during structural fire attacks?

By creating a protective water curtain and cooling hot gases.

Why is nozzle selection considered a strategic decision in firefighting?

Because it directly impacts suppression tactics, firefighter safety, and resource efficiency.

What should firefighters do to ensure effective nozzle use during operations?

Engage in regular training and drills to become proficient with each nozzle type and understand when to use them.

Chapter 3: Gallons Per Minute (GPM) vs. British Thermal Units (BTU)

Balancing Gallons Per Minute (GPM) and British Thermal Units (BTU) is the essence of effective fire suppression. In firefighting, these two measurements can make all the difference between success and failure. GPM represents the flow rate of water delivered to combat fires, while BTU measures the heat energy produced by a blaze. Understanding the delicate balance between these two factors is crucial for firefighters tackling intense fire scenarios. The relationship between water flow and heat output must be accurately assessed to ensure that the resources match the demands of various fire situations. This interplay forms the foundation upon which strategies are built, enabling firefighters to anticipate and respond effectively to different types of fires.

Throughout this chapter, you will explore the nuanced relationship between GPM and BTU, delving into their individual roles and how they interact within the context of firefighting. We will examine key concepts such as Heat Release Rate (HRR), a crucial metric that guides the allocation of resources during a firefight. You'll learn about the strategic calculations necessary for determining optimal GPM to counteract varying levels of BTU output and the real-world challenges faced in implementing these calculations. Additionally, we'll discuss advanced techniques for overcoming high-intensity fire loads, utilizing enhanced water flow rates, and employing additional suppression agents to bolster firefighting efforts. Our aim is to equip you with the knowledge needed to make informed decisions on the ground, ensuring both effectiveness and safety in every situation.

Introduction to GPM and BTU in Fire Suppression

IN THE REALM OF FIREFIGHTING, two main concepts that are pivotal to understanding and successfully combating fires are Gallons Per Minute (GPM) and British Thermal Units (BTU). These terms might seem straightforward, but they hold substantial significance in the world of fire suppression. To effectively address fires and control their spread, firefighters must comprehend the relationship between GPM and BTU. This understanding not only aids in deploying resources efficiently but also ensures that tactics align with the intensity of the fire at hand.

To begin with, let's explore GPM, which stands for Gallons Per Minute. GPM is a measure of water flow rate and is crucial in the context of firefighting because it represents how much water can be delivered to a fire scene in a given minute. When faced with the task of extinguishing a fire, water serves as the primary medium to cool down the flames and suppress the combustion process. Therefore, having an adequate GPM is vital since it determines whether sufficient water volume is being delivered to control or eventually extinguish the fire. In practical terms, higher GPM means more water is available to absorb the heat generated by the fire, thus playing a direct role in reducing the flame's intensity and helping prevent the fire from spreading to other areas.

On the other hand, we have BTU, short for British Thermal Units. BTU is a measurement of heat energy, defining how much energy is needed to raise the temperature of one pound of water by one degree Fahrenheit. In the context of fire, BTU indicates the amount of heat energy that a fire is producing. The higher the BTU, the more intense and potentially challenging the fire is to control. A blaze with high BTU output requires more extensive efforts to be subdued, as it emits greater amounts of heat energy into the environment, making it harder to tackle effectively without significant resources.

Now, understanding the interplay between GPM and BTU is paramount for firefighters when developing their strategies. The goal is to ensure that the water flow rate, represented by GPM, matches the heat energy produced by the fire, indicated by BTU. If a fire's BTU output surpasses the cooling power of the water being applied (determined by the GPM), then the efforts to suppress the fire will likely fall short. This misalignment could lead to the fire intensifying and possibly spreading uncontrollably. Conversely, if firefighters deploy a water flow rate that exceeds the fire's heat energy output, they stand a better chance of not only controlling the flame but ultimately extinguishing it.

For instance, imagine a scenario where a room filled with flammable materials catches fire, generating substantial heat. The firefighters on the scene need to quickly assess the BTU output of this blaze. Using their knowledge of the necessary GPM to counteract such intensity, they calculate how much water flow is required to douse the flames effectively. With the right GPM, firefighters can gain the upper hand, bringing the blaze under control before it spreads beyond the initial area.

The correlation between GPM and BTU is not merely mathematical; it's a strategic consideration that informs decisions in real-time situations. Firefighters evaluate the environment, the materials burning, and the potential for the fire to grow. Each of these factors contributes to estimating both the BTU output and the necessary GPM for effective intervention. Through this understanding, fire response teams can make informed choices about resource deployment, including how many hose lines are needed, the size of nozzles used, and the pressure settings for optimal water delivery.

Moreover, modern firefighting techniques often involve sophisticated equipment and strategic application of water. While GPM provides the basic measure of water flow, technologies such as variable flow nozzles allow adjustments in real time, catering to changing conditions during active firefighting operations. This flexibility further reinforces the importance of understanding the link between water flow rates and heat energy outputs – allowing firefighters to adapt their approach dynamically, ensuring efficiency and safety.

Flow Rate vs. Heat Release Rate

MATCHING WATER FLOW rates to fire intensity is a critical aspect of effective fire suppression. At the core of this concept lies the Heat Release Rate (HRR), an essential metric that quantifies how swiftly a fire emits heat energy. Understanding HRR is vital as it directly informs the required suppression response, ensuring firefighters can appropriately match their strategy and resources to the fire's intensity.

The HRR essentially acts like a pulse-check for the fire, giving insights into its size and potential destructiveness. For example, a small kitchen fire might have an HRR of about 200,000 BTU, while a large industrial fire could exceed several million BTU. This discrepancy in energy output necessitates varying approaches regarding water flow.

To effectively plan firefighting strategies, understanding how to calculate the necessary Gallons Per Minute (GPM) is crucial. The formula GPM = BTU / (Water's heat absorption efficiency × 8.34 lbs/gal × temperature rise) provides firefighters with a clear guideline on determining the correct amount of water flow required to combat specific fire scenarios successfully.

Let's break down the formula: The numerator, BTU, represents the fire's heat energy potential. The denominator factors in the efficiency with which water absorbs this heat, considering both its weight and the anticipated increase in temperature due to the fire.

Consider a residential fire with an HRR of 10,000,000 BTU. Using the GPM formula, you'll calculate the water flow needed based on available water properties and targeted temperature decrease. Suppose the efficiency of water in absorbing heat is about 0.5, with a desired temperature rise of 10°F. You can derive the necessary GPM to ensure effective firefighting. By plugging these values into the formula, firefighters can determine how much water flow is needed to suppress the fire efficiently and safely (Diasana, 2020).

On a larger scale, high-BTU scenarios, such as industrial fires, demand even greater precision and planning. Imagine an HRR of 50,000,000 BTU due to large quantities of flammable materials or chemicals. Here, the calculated GPM would be significantly higher, reflecting the intense energy release and increased threat level posed by such incidents. These cases underscore the importance of considering both HRR and appropriate GPM calculations to strategize effective responses.

In practice, however, implementing these calculations isn't without challenges. Pump capacity and water supply limitations play significant roles in determining whether calculated GPM solutions can be realized effectively. Firefighters must consider the capabilities of their equipment. A pump that can only handle minimal GPM will struggle in high-BTU scenarios, potentially necessitating additional reinforcements or alternative suppression methods (Amara, 2019).

Beyond individual firefighting units, municipal infrastructure also impacts practical application. Municipal water systems may not always provide sufficient pressure or volume to meet calculated GPM demands, particularly in older neighborhoods or rural areas. This means preemptively assessing local water supply capabilities through flow tests becomes imperative in planning and adjusting strategies accordingly.

For instance, if a community's hydrants can't deliver the needed volume, contingency plans involving portable tanks or drafting from alternative water sources such as nearby lakes or reservoirs must be put in place. Techniques like these ensure no matter the scenario, firefighters have access to the necessary resources to effectively suppress fires.

Overall, recognizing HRR as an indicator of fire intensity and leveraging the GPM formula provides invaluable guidance for matching water flow rates to the demands of any fire incident. Such strategic insight aids firefighters in optimizing resource allocation and executing precise suppression tactics tailored to specific conditions and constraints.

Overcoming High-BTU Fire Loads

IN THE REALM OF FIREFIGHTING, comprehending the dynamics between Gallons Per Minute (GPM) and British Thermal Units (BTU) is crucial for effectively managing fires with high heat energy output. One pivotal strategy involves crafting tailored approaches for handling scenarios characterized by significant fire loads, such as chemical fires or fires within large storage facilities. These scenarios demand more intensive suppression efforts due to their potential for rapid escalation and widespread damage.

To navigate high-BTU situations, it's essential to first identify environments where extreme fire loads are likely to occur. Chemical fires, for example, often generate substantial heat and present unique hazards due to volatile materials. Similarly, fires in large storage facilities can spread rapidly given the accumulation of combustibles. Understanding these conditions allows firefighters to anticipate the scale of response needed and adjust their tactics accordingly.

Strategically managing high-BTU fires necessitates employing enhanced water flow rates via increased pump capacity, a technique that can significantly impact the outcome of firefighting operations. By deploying pumps capable of delivering higher GPM, firefighters can ensure a more robust suppression effort that matches the intensity of the fire's thermal output. Discussing limitations such as pump capacity and water supply is vital here, as these factors directly influence the feasibility of sustaining such high water demands.

Moreover, the effective use of additional suppression agents like foams and chemicals plays a critical role in controlling fires with extreme BTU outputs. These agents can help suppress flames more efficiently than water alone, providing an additional layer of defense in tackling intense blazes. Implementing defensive tactics is another strategic approach worth considering, particularly in scenarios where aggressive offense might not be viable. Setting up barriers to contain the fire or prioritize protecting certain structures over others could minimize damage while enhancing firefighter safety.

Recounting both successful and unsuccessful attempts at high-BTU fire suppression offers invaluable insights for evolving strategies. Historical case studies serve as learning tools, highlighting adaptive tactics and common pitfalls. For instance, in a notable incident involving a chemical warehouse fire, the quick deployment of foam blankets alongside traditional water streams significantly curtailed the fire's progression, showcasing the effectiveness of combining suppression techniques. Conversely, cases where inadequate assessment of fire load led to resource misallocation underscore the importance of thorough preparedness.

Accurate fire size-up is a cornerstone of preparedness when confronting high-BTU scenarios. This involves quickly yet meticulously assessing the extent of the fire, potential growth trajectories, and available resources on site. Highlighting the importance of accurate fire size-up to estimate the Heat Release Rate (HRR) ensures that firefighters have a clear understanding of the energy involved and can allocate resources appropriately. A systematic evaluation allows for more informed decisions regarding deployment of equipment and personnel, reducing the likelihood of being overwhelmed by unexpectedly intense fires.

Resource allocation tailored specifically to potential high-BTU fires further underscores the need for meticulous strategy development. Establishing preemptive measures, such as positioning reserve equipment or coordinating with neighboring departments for mutual aid, enhances readiness when facing formidable fire challenges. Ensuring that teams are well-versed in deploying varying suppression agents and adjusting tactics based on real-time assessments cultivates a more agile firefighting force capable of adapting to diverse high-BTU incidents.

Ultimately, developing strategies for high-BTU scenarios is about merging technical expertise with practical experience. It requires firefighters to not only understand the quantitative aspects of GPM and BTU but also apply this knowledge dynamically in the field. By identifying high-risk scenarios, exploring strategic approaches, learning from past experiences, and emphasizing precise fire size-ups and resource allocation, firefighting teams can enhance their ability to manage significant fire threats effectively.

Final Thoughts

UNDERSTANDING HOW GALLONS Per Minute (GPM) and British Thermal Units (BTU) correlate is essential for firefighters aiming to optimize their fire suppression tactics. This chapter has delved into the distinct roles of GPM as a measure of water flow rate and BTU as an indicator of heat energy, emphasizing the importance of aligning these two elements in firefighting scenarios. By exploring formulas and methods to calculate the necessary GPM based on a fire's BTU output, firefighters are equipped with the knowledge to craft effective strategies for handling fires of varying intensities. This understanding ensures that resources are utilized efficiently, aligning water delivery with the demands posed by different fire situations.

In addition to theoretical insights, practical considerations such as pump capacity, municipal water supply limitations, and high-BTU environments have been highlighted as critical factors influencing real-world applications. The narrative underscores the need for accurate assessments and adaptive approaches tailored to specific fire conditions. Leveraging modern techniques and equipment, including variable flow nozzles and supplementary agents like foams, further enhances the capability to manage intense blazes effectively. By integrating both quantitative assessments and strategic flexibility, this chapter provides a comprehensive framework for improving preparedness and response in challenging firefighting operations, equipping firefighters to face diverse fire threats with confidence and precision.

What does GPM stand for, and why is it important in firefighting?

GPM stands for *Gallons Per Minute*, which measures the water flow rate. It is vital in firefighting because it indicates how much water is delivered to suppress a fire, directly impacting the ability to absorb heat and control the blaze.

What is a BTU, and how is it relevant to fire suppression?

BTU stands for *British Thermal Unit*, a measurement of heat energy. It reflects how much heat a fire produces, which helps firefighters determine the intensity and the resources needed for suppression

How are GPM and BTU related in firefighting operations?

GPM and BTU are directly related; the water flow rate (GPM) must be sufficient to absorb the heat energy (BTU) produced by a fire. Proper balance ensures effective suppression.

What happens if GPM is too low for the fire's BTU output?

If GPM is too low, the water won't absorb enough heat, and the fire may grow or spread uncontrollably, endangering property and lives.

Define Heat Release Rate (HRR).

HRR is the rate at which a fire releases heat energy, measured in BTU. It helps determine the scale of response needed to suppress the fire effectively.

What formula is used to calculate required GPM for a fire?

GPM = BTU / (Efficiency × 8.34 × Temperature Rise)

This formula calculates the water flow needed based on fire intensity and desired temperature change.

In the GPM formula, what does 8.34 represent?

8.34 is the weight in pounds of one gallon of water, which helps quantify how much energy water can absorb per degree Fahrenheit.

What is the typical efficiency factor used in GPM calculations?

A typical water heat absorption efficiency factor is **0.5**, accounting for water loss and imperfect heat transfer during firefighting.

Why is understanding HRR essential in high-BTU fire scenarios?

It allows firefighters to estimate fire intensity quickly and adjust suppression tactics and resource deployment to avoid being overwhelmed.

Give an example of a high-BTU fire scenario.

Fires in chemical warehouses or large storage facilities are high-BTU scenarios due to the large volume of combustible materials and potential for intense heat output.

What role does pump capacity play in GPM delivery?

Pump capacity limits how much water can be delivered. In high-BTU situations, inadequate pump capacity may prevent achieving the necessary GPM.

What are potential municipal challenges when meeting high GPM needs?

Limited water pressure or flow from hydrants, especially in older or rural systems, can restrict water availability, requiring alternative sources.

Name two alternate water supply methods for high-demand incidents.

1. Portable water tanks

2. Drafting from natural water sources like lakes or rivers

How do variable flow nozzles assist in firefighting?

They allow firefighters to adjust water flow rates in real-time, matching changing fire intensities without needing equipment changes.

What additional suppression agents are often used for high-BTU fires?

Firefighting foams and chemical agents, which help smother flames and cool surfaces faster than water alone.

What is a defensive tactic in high-BTU fire scenarios?

Creating barriers to contain the fire or focusing efforts on protecting exposures rather than direct suppression of the main fire.

How does fire size-up support better GPM and BTU balancing?

Accurate fire size-up helps estimate HRR, allowing for better planning of suppression strategies and resource deployment.

What is the risk of underestimating a fire's BTU output?

It can lead to insufficient water flow, delayed control efforts, increased damage, and higher danger to firefighters.

Why is real-time adaptability crucial during firefighting operations?

Fire conditions change rapidly; the ability to adjust water flow, suppression agents, and tactics ensures safety and effectiveness.

What's the overall strategic takeaway from understanding GPM vs. BTU?

Firefighters must match water delivery to heat energy output, considering equipment, environment, and fire behavior, to suppress fires effectively and safely.

Chapter 4: Hose Size and Application

Selecting the right hose size for firefighting operations is essential to ensuring maximum efficiency and effectiveness in the field. Each firefighting mission depends on choosing hoses that meet specific requirements, tailored to the demands of various scenarios. This decision involves comprehending not only the physical characteristics of the hoses but also how different aspects like diameter, length, and water pressure can affect performance during suppression activities. The intricacies of this process involve understanding how these elements impact the ability to control fire efficiently, emphasizing the importance of knowing one's equipment inside out. Capturing this essence allows firefighters to prepare adequately for any situation, maintaining readiness and adaptability at all times.

In this chapter, you'll explore the critical factors affecting hose selection, focusing on how diameter and length influence handling and efficiency. Examining friction loss offers insights into how water moves through hoses, revealing the balance needed between hose size and pressure for optimal performance. You'll uncover the significance of nozzle compatibility, connecting it with hose choices to maximize water delivery while minimizing losses. Dive into practical scenarios illustrating the benefits and limitations of using specific hose sizes and learn strategies to overcome challenges posed by varying environments and conditions. By the end of this chapter, you'll be equipped with the knowledge necessary to make informed decisions when selecting the most suitable hoses for different firefighting tasks, enhancing your overall operational effectiveness.

Introduction to Hose Selection and Diameter

IN FIREFIGHTING, HOSE diameter plays a pivotal role in the effectiveness of fire suppression efforts. With various types of hoses available, it is crucial to select the right size for specific tasks to ensure that the response is as efficient and effective as possible. The importance of hose diameter becomes evident when considering different firefighting scenarios, where each situation demands a particular approach to water delivery.

Hoses are indispensable tools for firefighters, serving multiple purposes from attacking the fire directly to supplying water over long distances. Choosing the correct diameter involves understanding the task-specific needs of each operation. For instance, a small diameter hose, typically 1¾ inches, is highly favored for interior fire attacks. Its compact size and lightweight nature allow firefighters to navigate through tight spaces such as hallways and stairwells with ease. This smaller hose can be maneuvered more swiftly, lending itself to quick deployment, which is critical in residential or vehicle fires where rapid suppression is required.

The agility provided by a 1¾-inch hose does not come at the expense of performance. While it delivers lower volumes of water compared to larger hoses, it operates at higher pressures, achieving flow rates between 140 and 200 gallons per minute (gpm). These attributes make it an all-around choice for initial fire attack lines, especially in situations demanding speed and precision. However, its limitation lies in larger fire incidents, where higher water flows are necessary to combat the increased heat release rate effectively. In such instances, reliance on this small diameter could hinder overall suppression efforts if used inappropriately. (Hose, 2021)

Transitioning to medium diameter hoses, the 2½-inch size emerges as a versatile option, bridging the gap between mobility and ample water delivery. This hose serves crucial roles in both interior firefighting within larger structures and exterior operations requiring sizeable streams. It can deliver water at flow rates ranging from 200 to 300 gpm, providing a substantial amount of water to tackle severe blazes in commercial buildings or during exposure protection. Although heavier and more challenging to handle than its smaller counterpart, the balance it strikes between capacity and maneuverability cannot be understated.

Firefighters often deploy 2½-inch hoses when confronting fires that demand high-flow operations while still necessitating some level of movement around the fireground. Handling these hoses generally requires a team of three to four firefighters, emphasizing the importance of teamwork and coordination, especially in confined environments or when repositioning the line quickly is vital. Despite its enhanced capabilities, like any tool, this hose's efficiency depends on the specific application and conditions it encounters, such as entry points, fire load, and structural layout. (Hose, 2021)

For scenarios involving extensive water supply demands, large diameter hoses, specifically those measuring 4 or 5 inches, become the go-to choice. These supply lines are integral in connecting hydrants to fire engines, ensuring a steady and reliable water source. By leveraging their broad diameters, these hoses minimize friction loss, allowing massive volumes of water to journey over significant distances at lower pressures. As a result, they play an essential role not only in sustaining prolonged firefighting activities but also in supporting operations that require drafting water from distant sources or feeding aerial devices.

However, deploying large diameter hoses introduces challenges akin to their benefits. Their sheer size and weight complicate maneuvers once charged with water, necessitating careful consideration of deployment strategy and staffing. Often, it takes a crew of several firefighters to lay out and charge these hoses effectively, highlighting the need for adequate training and planning in advance. Moreover, understanding the influence of diameter on hydraulic performance is imperative since kinks or twists in the hose can reduce efficiency, impacting the overall water flow and pressure at the nozzle. Fireground commanders must thus evaluate terrain and operational needs meticulously to prevent unnecessary obstructions and ensure optimal functionality. (Isakson, 2024)

Ultimately, recognizing the critical role played by hose diameter in firefighting success means appreciating how each type brings distinct advantages and limitations. From flexible 1¾-inch attack lines suitable for rapid intervention to robust 5-inch supply hoses underpinning structured operations, harnessing their potential demands insight into both technical specifications and situational requirements. Fire services continually strive towards refining their equipment choices, informed by evolving technology and field experiences that enhance strategic decision-making, ensuring readiness for increasingly complex fire scenarios.

Understanding and Mitigating Friction Loss

FRICTION LOSS IS AN important concept when discussing the effective application of hoses in firefighting. It refers to the unavoidable decrease in water pressure as it flows through a hose, primarily influenced by the diameter and length of the hose. Understanding friction loss is crucial, as it directly impacts the delivery and efficiency of water to a fire scene.

To grasp how friction loss works, consider this: When water moves through a smaller diameter hose, there is more contact with the hose walls, creating greater resistance and thus higher friction loss. Conversely, larger hoses offer less resistance and maintain more pressure, allowing for more efficient water flow. Additionally, the longer the hose, the more significant the pressure drop due to the increased distance the water must travel, which amplifies the friction encountered along the way.

An example can further illustrate this phenomenon. Suppose a team uses a 1¾-inch hose for a fire attack inside a building. The crew may notice a substantial drop in pressure if the hose length extends beyond 200 feet. Switching to a 2½-inch hose might reduce friction loss, maintaining the necessary pressure for effective firefighting, especially in high-rise scenarios or where extensive lengths are unavoidable. By calculating friction loss beforehand, firefighters can anticipate pressure requirements and adjust their setup accordingly.

Mitigating friction loss involves using strategies that focus on maximizing water delivery with minimal pressure loss. One effective technique is utilizing larger diameter hoses wherever feasible. A 4-inch or 5-inch supply line, for instance, significantly reduces friction loss compared to smaller hoses, providing a steady and reliable water supply even over long distances. This approach is particularly useful in scenarios requiring massive water flow, such as industrial fires or large structural incidents.

Another key strategy involves minimizing sharp bends in hose layouts. Each bend introduces additional friction, reducing pressure and flow rates. Whenever possible, firefighters should aim for straight hose lines. If turns are necessary, gentle curves can help maintain smoother water movement and preserve pressure integrity. Being mindful of hose routing during deployment is vital to prevent unnecessary friction points and ensure optimal performance.

The role of pump operation is instrumental in managing friction loss effectively. Pumps compensate for pressure drops by adjusting output based on calculated friction losses for given hose configurations. Firefighting teams need to be adept at operating pumps, fine-tuning pressures to match water delivery needs on the ground. Regular training on pump operations and friction loss calculations enhances the team's ability to adapt quickly to different situations, ensuring they keep water flowing at the required pressure.

Firefighters must also acknowledge the interplay between friction loss and the nozzle choice. Though this is discussed elsewhere in the chapter, understanding how different nozzles affect water flow helps in aligning equipment to minimize overall pressure loss. Coordinating nozzle selection with hose diameter and anticipated friction loss contributes to maintaining effectiveness during fire operations.

Challenges arise, particularly when deploying large-diameter hoses. These hoses, while effective in reducing friction loss, can be cumbersome to maneuver. Ensuring the team is well-versed in best practices for handling such hoses minimizes operational delays and maximizes their benefits.

For instance, deploying these hoses requires careful coordination, clear communication, and sometimes mechanical aids like hose reels to expedite the setup process. Following proper protocols during deployment ensures the hoses are adequately laid out, avoiding kinks and unnecessary turns that could hinder performance. Moreover, regular maintenance checks and training exercises on handling large-diameter hoses are essential components in overcoming deployment challenges.

Effective Hose and Nozzle Pairing

MATCHING HOSES TO APPROPRIATE nozzles is a critical aspect of firefighting that can significantly influence the effectiveness of an emergency response. Properly aligned hose-nozzle combinations ensure that firefighters maintain optimal flow rates and pressure levels, which are essential for effective fire suppression. By understanding the relationship between hose size and nozzle type, departments can enhance their operational outcomes and reduce potential risks. This section will explore why correct matching is important, the types of nozzles available, practical examples of effective combinations, and common troubleshooting scenarios.

Ensuring optimal performance in firefighting operations begins with selecting the right hose and nozzle combination. When matched correctly, these components work together to achieve the desired flow rate and pressure levels, allowing the firefighting team to carry out efficient extinguishing operations. An improperly matched combination might result in inadequate water delivery or excessive pressure, potentially hindering firefighting efforts or compromising safety. For instance, if a nozzle designed for low-pressure operation is paired with a high-flow-rate hose, this can lead to uncontrolled water spray, making it difficult to control the fire effectively.

There are several types of nozzles used in firefighting, each requiring specific hose sizes to maximize efficiency. The fog nozzle is designed to produce a fine spray, creating a heat-absorbing mist that is highly effective for cooling superheated environments. For fog nozzles, matching them with a hose that allows for the appropriate flow rate without excessive friction loss is crucial. A smooth bore nozzle, on the other hand, delivers a solid stream of water ideal for penetrating deep into the fire's base. This type requires a sufficiently large hose to provide the necessary water volume and pressure. Lastly, combination nozzles, which offer both fog and straight stream options, need hoses that can support variable flow rates while maintaining consistent pressure.

Practical examples illustrate how effective combinations of hoses and nozzles yield successful firefighting outcomes. Imagine a scenario where a 2½-inch hose is paired with a smooth bore nozzle for a commercial building fire. This setup would allow the crew to deliver substantial water volumes at lower pressures, ideal for larger fires that demand high flow rates. Contrarily, using a 1¾-inch hose with a fog nozzle might be more appropriate for confined spaces such as residential interiors, where maneuverability and precision are vital. Such strategic pairing not only enhances the firefighting capability but also mitigates potential hazards like excessive nozzle reaction, ensuring safer and more controlled operations.

Addressing common troubleshooting scenarios can greatly enhance overall firefighting efficacy and response time. One frequent issue is kinking in the hoseline, which can restrict water flow and reduce pressure at the nozzle. This problem often arises when a hose with an inappropriate diameter is used. Ensuring that the hose is well-suited for its intended flow rate can minimize such occurrences. Additionally, teams may encounter challenges with nozzle compatibility, particularly when using modern low-pressure nozzles with older hose models. Training sessions and regular equipment checks can help preemptively identify and rectify these mismatches, keeping fire crews prepared and well-equipped.

By focusing on proper hose and nozzle matchups, fire departments can improve their tactical operations significantly. The implications of these choices extend beyond immediate firefighting efforts, influencing long-term maintenance costs, training programs, and overall readiness. Well-matched equipment reduces wear and tear, lowering replacement and repair expenses. Moreover, firefighters who regularly train with optimally matched gear develop greater confidence and proficiency, further enhancing their performance during emergencies.

Training is an indispensable element of ensuring that equipment is used correctly and efficiently. Providing firefighters with hands-on experience allows them to understand the nuances of different nozzle-hose combinations and appreciate the impact of each setup in various scenarios. Regular drills that mimic real-life situations enable them to quickly assess the best equipment pairings under pressure, minimizing decision-making delays during actual incidents. Training should emphasize adapting to different fireground environments and conditions, reinforcing the importance of flexibility in their approach.

Ultimately, the goal of hose-nozzle matching is about ensuring that every component of the firefighting system works in harmony to deliver the needed results. Departments must prioritize understanding the specific needs of their jurisdiction, taking into account factors like typical building structures, common fire types, and resource availability. By doing so, they can make informed decisions that optimize their equipment usage and strategic deployments.

Summary and Reflections

IN THIS CHAPTER, WE'VE delved into the intricacies of hose selection, exploring how crucial factors like diameter, length, and application influence firefighting efficiency. Understanding these elements helps firefighters choose the right tool for the task at hand, whether maneuvering a 1¾-inch hose in tight residential spaces or deploying large diameter hoses to supply water over long distances. Each hose type comes with its unique advantages and potential challenges, such as handling and friction loss, which require strategic consideration to optimize their use on the fireground.

By aligning hose types with appropriate nozzles, firefighters can enhance water delivery and maintain control during emergency responses. Familiarity with friction loss calculations ensures that crews anticipate pressure needs and make informed decisions about hose configurations. Keeping these principles in mind, departments can improve operational success by continually refining equipment choices and adapting to new technologies and field experiences. Ultimately, mastering hose selection is about empowering firefighters with the knowledge to handle diverse scenarios effectively, keeping themselves safe while combating blazes efficiently.

Chapter 4 Questions and Answers

Why is selecting the correct hose size crucial for firefighting? The correct hose size ensures efficient water delivery, maintains pressure, and supports the firefighter's ability to control the fire effectively, based on the specific operational needs.

What is the primary benefit of a 1¾-inch hose?

The 1¾-inch hose is favored for its compact size, lightweight, and high pressure, making it ideal for quick deployment in interior fire attacks, especially in tight spaces like hallways and stairwells.

What are the flow rates typically achieved by a 1¾-inch hose?

The flow rate for a 1¾-inch hose typically ranges between 140 and 200 gallons per minute (gpm).

When is a 2½-inch hose most commonly used?

The 2½-inch hose is used for both interior firefighting in larger structures and exterior operations requiring high-flow operations, such as in commercial buildings or during exposure protection.

What are the challenges associated with a 2½-inch hose?

The 2½-inch hose is heavier and more difficult to handle than smaller hoses, requiring a team of three to four firefighters for effective operation, especially in confined environments.

Why is a 4 or 5-inch hose used in firefighting?

A 4 or 5-inch hose is used as a supply line to connect hydrants to fire engines, delivering large volumes of water with minimal friction loss, especially over long distances or when drafting water.

What are the challenges of using large diameter hoses (4 or 5 inches)?

Large diameter hoses are heavy and cumbersome to maneuver, requiring several firefighters to deploy and manage effectively, as well as careful consideration of deployment strategy and staffing.

How does hose diameter affect friction loss?

Smaller diameter hoses cause higher friction loss due to greater contact with the hose walls, while larger hoses experience less friction loss, allowing for more efficient water flow and higher pressure.

How can firefighters mitigate friction loss?

Firefighters can mitigate friction loss by using larger diameter hoses, minimizing sharp bends in hose layouts, and ensuring proper pump operation to compensate for pressure drops.

What impact does hose length have on friction loss?

Longer hoses experience greater friction loss due to the increased distance water must travel, which leads to a more significant pressure drop.

How does nozzle choice influence friction loss?

Nozzles that are mismatched with hose size or flow rates can exacerbate friction loss, affecting water flow and pressure, which can hinder firefighting efforts.

What is the role of pump operation in managing friction loss?

Pumps help manage friction loss by adjusting output based on the calculated friction loss of the hose configuration, ensuring adequate water pressure at the nozzle.

What is the effect of sharp bends in a hose line?

Sharp bends in a hose line increase friction loss and reduce water pressure and flow rates, making it harder to maintain effective fire suppression.

Why is it important to understand hose-nozzle compatibility?

Correct hose-nozzle compatibility ensures that the firefighting team delivers the optimal flow rate and pressure, preventing inadequate water delivery or excessive pressure that could compromise the operation.

How does a smooth bore nozzle differ from a fog nozzle?

A smooth bore nozzle delivers a solid stream of water for deep penetration into the fire's base, while a fog nozzle produces a fine mist ideal for cooling and heat absorption in hot environments.

What is the ideal nozzle and hose combination for a commercial building fire?

A 2½-inch hose paired with a smooth bore nozzle is ideal for commercial building fires, allowing substantial water volumes at lower pressures for larger fires.

How does a 1¾-inch hose and fog nozzle combination benefit residential fires?

A 1¾-inch hose with a fog nozzle is perfect for residential fires, offering maneuverability in confined spaces while providing cooling and heat absorption with the fine mist of the fog nozzle.

What common issue can occur when a hose is mismatched with a nozzle?

A common issue is excessive water spray or inadequate water delivery, leading to inefficient firefighting or compromised safety.

How can fire departments reduce maintenance costs related to hose-nozzle systems?

By regularly training firefighters on hose-nozzle compatibility, performing equipment checks, and using appropriately matched gear, fire departments can reduce wear and tear, leading to lower maintenance and replacement costs.

What is the importance of training for effective hose and nozzle pairing?

Regular training ensures firefighters understand the nuances of hose and nozzle combinations, allowing them to select the best setup under pressure during real-life emergencies, improving operational efficiency and safety.

Chapter 5: Types of Fire Department Appliances and Tools

Firefighting involves far more than just extinguishing flames; it's about mastering the myriad appliances and tools that make battling fires possible. From towering high-rise buildings to sprawling industrial sites, firefighters must be prepared for any environment or scenario they may encounter. This preparation often hinges on their knowledge and expertise in using a vast array of equipment designed to control and eliminate fire hazards effectively. The evolution of these appliances and tools showcases innovative approaches to overcoming the unpredictable nature of firefighting. Whether it's delivering water to the top floors of skyscrapers or penetrating walls to reach hidden fires, the right equipment can make all the difference. By understanding each tool's unique functionality, firefighters enhance their ability to address emergencies efficiently and safely.

This chapter delves into the essential tools and appliances crucial for modern firefighting operations. It unpacks various fire suppression systems like standpipes and monitors, highlighting how they equip teams to manage large-scale blazes and structural threats. Readers will explore the tactical use of foam systems, which are tailored for specialized fire scenarios, and learn how rescue tools such as Halligan bars and thermal imaging cameras prove indispensable in challenging conditions. Furthermore, the importance of efficient water management and tools like hose clamps and hydrant wrenches is emphasized, illustrating how every piece of equipment plays a pivotal role in successful fire response efforts. With detailed insights into both standard and specialized gear, this chapter equips firefighters with the knowledge needed to wield their instruments with precision and confidence, ultimately enhancing their life-saving capabilities in the field.

Fire Suppression Appliances

FIRE DEPARTMENTS AROUND the world rely on a variety of essential appliances and tools to effectively suppress fires and conduct rescue operations. Understanding the key appliances used in fire suppression is crucial for firefighters, as these devices play a vital role in managing different types of fire scenarios.

One of the primary systems used in high-rise firefighting is the standpipe system. These are critical for delivering sufficient water supply to upper floors during emergencies. Standpipe systems come in three main types: wet, dry, and combination. Wet standpipe systems constantly maintain water in the pipes, ensuring immediate availability. This type is particularly advantageous in buildings susceptible to severe blaze outbreaks. Conversely, dry standpipe systems do not hold water until needed, reducing the risk of water damage from broken pipes during colder weather. Combination systems integrate the benefits of both by providing pressurized water when activated while remaining dry during non-emergency periods (*Fire Suppression - Monitors & Appliances*, 2024). Each type has unique operational advantages, allowing flexibility in diverse firefighting situations.

Monitors are another indispensable appliance. They are responsible for delivering high-volume water streams over wide areas, which is especially useful in large-area fires. These can be either fixed or portable. Fixed monitors are strategically positioned on fire trucks or at specific locations within industrial sites, offering sustained suppression capabilities. Portable monitors provide versatility; they can be placed wherever necessary, depending on the fire's behavior and progression. However, effective use of monitors requires careful consideration of their positioning and flow control to maximize coverage without wasting resources (*Firefighting Strategies and Tactics - Third Edition. Flashcards*, 2024).

Deck guns are a staple in fire suppression due to their ability to rapidly deploy substantial volumes of water. Mounted on fire apparatuses, deck guns can swivel and adjust elevation, allowing firefighters to target flames with precision. These guns are especially valuable in situations demanding quick effort to extinguish fires before they escalate. Their capability to deliver powerful streams makes them ideal for controlling blazes that could otherwise spread uncontrollably, such as those encountered in industrial fires or large outdoor incidents (*Fire Suppression - Monitors & Appliances*, 2024).

Foam systems represent specialized applications in firefighting, designed to combat flammable liquid fires and deep-seated structural blazes. There are various types of foam systems tailored for distinct challenges: Class A foams enhance water's effectiveness against ordinary combustibles like wood and paper, producing thick blankets that suppress oxygen. Class B foams, on the other hand, are formulated for flammable liquids, helping prevent re-ignition by forming barriers between the fuel and air. Compressed Air Foam Systems (CAFS) further extend these capabilities, providing an efficient distribution mechanism that improves reach and adhesion. By understanding how to deploy these foam systems effectively, firefighters can better manage complex fire scenarios that involve hazardous materials (*Firefighting Strategies and Tactics - Third Edition. Flashcards*, 2024).

Essential Tools

FIREFIGHTING DEMANDS the use of specialized tools, each crafted to meet the challenges faced in suppression and rescue operations. Among these, the Halligan bar stands out as a quintessential tool for forcible entry. This versatile implement features a fork, blade, and adze for prying open doors or windows, making it indispensable during urgent rescues. When paired with an axe, either pickhead or flathead, firefighters can maximize their entry capabilities. The pickhead axe is particularly useful for breaching walls or roofs, while the flathead variety serves effectively as a hammer against the Halligan, forming a dynamic duo essential in high-pressure situations.

Thermal imaging cameras (TICs) have become vital assets in modern firefighting. These devices allow firefighters to see through smoke and darkness, pinpointing hidden victims or hot spots not visible to the naked eye. In structural fires, TICs guide first responders toward trapped individuals by detecting body heat, ensuring no one is left behind. Their utility extends to wildland firefighting, where identifying and containing fire lines can prevent catastrophic spread. By providing a clear visual of the fire's dynamics, TICs also aid in strategic decision-making, significantly enhancing safety and effectiveness during operations.

Efficient water management remains critical in firefighting, underscoring the importance of hose clamps. These tools regulate water flow, allowing crews to make rapid adjustments without disconnecting hoses, thus preventing wastage and ensuring the right amount of water reaches the fire. Spanner wrenches facilitate the tightening and loosening of hose couplings, crucial for maintaining secure connections, while hydrant wrenches enable quick operation of fire hydrants. Together, they ensure swift water access, which is often a decisive factor in containment efforts.

Furthermore, adapters play a pivotal role by ensuring compatibility among various hoses and appliances. Fire departments may encounter different brands or types of equipment, but with the right adapters, seamless integration becomes possible. This adaptability not only saves precious time but also increases efficiency during responses, especially in mutual aid scenarios involving multiple agencies.

In addition to these fundamental tools, strainers and jet siphons provide essential support in drafting operations from static water sources. Strainers prevent debris from entering pumps, safeguarding their functionality, while jet siphons facilitate quick water transfer between tanks, maximizing available resources on site. Both components are integral when traditional water supply methods may be compromised, such as in rural or remote areas.

The strategic use of valves, including those from brands like Harrington, Akron, and Elkhart, further emphasizes control during firefighting endeavors. Valves are key in managing water direction and pressure, enabling firefighters to adapt quickly to changing conditions. Whether controlling flow into a portable water tank or regulating output from a pumper truck, their reliability is paramount.

Forcible entry tools extend beyond the Halligan and axes, encompassing bolt and wire cutters, which are invaluable for removing obstructions such as chains or padlocks that restrict access. Brooms, shovels, and rakes facilitate post-fire cleanup, helping to restore order and assess damages safely. Hooks and poles, including drywall hooks or pike poles, assist in ventilation and overhaul tasks, crucial steps in extinguishing residual flames and securing structures post-incident.

Saws, particularly those designed for wildland applications, address unique challenges posed by natural environments. Equipped to cut through thick vegetation, they aid in creating firebreaks or clearing pathways, enhancing the safety of both civilians and responders.

Wrenches from reputable suppliers ensure precision and durability in opening or closing a wide array of fittings. Brands like 5.11 Tactical, Elkhart Brass, and Akron Brass consistently equip departments with high-quality gear. Their commitment to innovation has led to tools specifically tailored for the rigorous demands of firefighting.

Acknowledging the diversity of fire tools and appliances illustrates the complexity of firefighting work. Each tool represents a vital component in the greater ecosystem of emergency response, and understanding their functions empowers firefighters to execute their duties with confidence and efficiency. Through continuous learning and adaptation, the fire service evolves, meeting the ever-changing landscape of hazards with resilience and dedication.

Specialized Equipment

IN FIREFIGHTING, SPECIALIZED equipment is vital for addressing unique challenges that arise during suppression and rescue operations. These tools enhance the effectiveness of fire departments in diverse situations, from rural settings to confined spaces requiring precise intervention or extrication efforts.

Portable pumps play a crucial role in ensuring water supply in remote or rural areas where access to hydrants may not be feasible. Lightweight and high-pressure variants are especially beneficial, providing the necessary mobility and power to reach fires in difficult terrains. One important aspect of utilizing portable pumps effectively involves their proper setup and maintenance. Firefighters need to ensure that these pumps are positioned correctly to maximize water flow while minimizing environmental impact. For example, placing the pump on stable ground prevents tilting that might disrupt water output. Regular maintenance checks, including inspecting hoses and connections for leaks, help prevent malfunctions in critical moments. Proper care and understanding of this equipment are essential to guarantee operational readiness when reaching fires far from traditional water sources. Tips for setup and maintenance, such as verifying fuel levels or running routine checks prior to deployment, can significantly boost efficiency in emergency scenarios (Keeping Industrial Zones Fire Safe – Know What to Keep in Check to Avoid Fire Hazards – Fire Fighting and Rescue Equipments, 2024).

Piercing nozzles offer a unique approach to battling fires trapped behind barriers. These nozzles feature designs that enable them to penetrate walls, roofs, or other obstacles, delivering water directly to concealed flames. This capability is particularly valuable in building collapses or attic fires where direct access is hindered. The correct use of piercing nozzles demands precise deployment techniques. Firefighters must assess structural integrity and choose the optimal angle and location for penetration to ensure maximum coverage within void spaces. Training in deploying these nozzles safely and efficiently is key to their success, as improper handling could compromise structural stability or lead to ineffective fire suppression. Practicing various deployment scenarios enhances proficiency and helps firefighters adapt swiftly to different structural challenges. Techniques for using these tools in tandem with other equipment, such as thermal imaging cameras, allow teams to locate hidden flames accurately before deploying the nozzle, ensuring swift and effective intervention (All Products, 2024).

Cutting tools form an integral component of firefighting equipment, providing essential support in extrication, ventilation, and creating safe access routes. Power saws, bolt cutters, and hydraulic cutters each have distinct applications, yet all require adherence to rigorous safety protocols due to their inherent risks. Power saws aid in cutting through walls, doors, and debris during rescue operations, facilitating quicker access to victims trapped inside buildings. Bolt cutters prove invaluable in slicing through locks or chains hindering entry, while hydraulic cutters possess the strength to sever reinforced materials commonly found in vehicles during extrications.

Using these cutting tools demands comprehensive understanding and training to operate them safely without endangering personnel or victims. Safety protocols include wearing appropriate personal protective equipment, maintaining safe distances from cutting zones, and conducting regular equipment inspections to ensure functionality. Additionally, teams must coordinate their efforts, communicating effectively to avoid accidental injuries during tool operation.

In extrication scenarios, the timely deployment of cutting tools can mean the difference between life and death. Consideration of the most efficient tool for the situation, coupled with strategic planning, allows for streamlined and safe rescue operations. The integration of cutting tools into well-practiced rescue procedures empowers teams to tackle complex emergencies with confidence. By continually updating training and incorporating lessons learned from real-world experiences, fire departments can improve their capabilities in utilizing cutting tools effectively while upholding the highest safety standards.

Main Points Recap

IN THIS CHAPTER, WE explored the vast array of essential equipment used in firefighting and rescue operations. We've delved into the critical role of standpipe systems, monitors, deck guns, and foam systems in effectively suppressing fires of various natures. From delivering water to upper floors with standpipes to targeting large-scale blazes with portable or fixed monitors, firefighters must understand each tool's unique advantages. Additionally, the importance of foam systems in battling both structural and flammable liquid fires underscores the need for strategic deployment in complex situations. Recognizing these appliances' applications allows firefighters to approach each scenario with the confidence that they have the right tools for effective fire management.

Beyond appliances, we've discussed specialized tools pivotal for rescue missions. The Halligan bar, thermal imaging cameras, and hose clamps highlight the diverse capabilities required to overcome entry, visibility, and water management challenges. Adaptors, strainers, jet siphons, and valves showcase how interoperability and efficiency are achieved through precise equipment coordination, particularly in multi-agency responses. Meanwhile, cutting tools ensure quick access in extrication scenarios, demanding ongoing training to maintain safety and precision. By mastering the use of these essential and specialized tools, firefighters enhance their ability to save lives and properties, equipping them to meet the demands of any emergency with preparedness and skill.

What are the three types of standpipe systems commonly used in high-rise firefighting?

Wet, dry, and combination standpipe systems.

What is the advantage of a wet standpipe system?

A wet standpipe system maintains water in the pipes at all times, ensuring immediate water availability during emergencies.

Why is a dry standpipe system useful in colder climates?

A dry standpipe system does not hold water until needed, reducing the risk of water damage from frozen pipes during colder weather.

How does a combination standpipe system function?

A combination system provides pressurized water when activated but remains dry during non-emergency periods.

What is the primary function of a fire monitor in firefighting?

A fire monitor delivers high-volume water streams over wide areas to suppress large fires effectively.

What is the difference between fixed and portable fire monitors?

Fixed monitors are strategically placed on fire trucks or industrial sites, while portable monitors can be moved and placed wherever needed for maximum suppression efficiency.

What is a deck gun and how is it used in fire suppression?

A deck gun is mounted on fire apparatus and can swivel and adjust elevation, allowing firefighters to target flames with precision and deliver large volumes of water quickly.

When is foam particularly useful in firefighting?

Foam systems are especially effective for combating flammable liquid fires and deep-seated structural blazes.

What are the two main types of foam systems used in firefighting?

Class A foam for ordinary combustibles and Class B foam for flammable liquids.

What is a Compressed Air Foam System (CAFS) and why is it advantageous?

CAFS provides efficient foam distribution, improving reach and adhesion, which is beneficial in complex fire scenarios.

What tool is essential for forcible entry during fire and rescue operations?

The Halligan bar is a versatile tool used for prying open doors and windows during forcible entry.

How does a thermal imaging camera (TIC) aid firefighters?

A TIC allows firefighters to see through smoke and darkness, helping them locate hidden victims and fire hotspots.

What is the purpose of hose clamps in firefighting operations?

Hose clamps regulate water flow, allowing crews to make quick adjustments without disconnecting hoses, ensuring efficient water delivery.

How do spanner wrenches assist firefighters in managing water supply?

Spanner wrenches help tighten and loosen hose couplings, ensuring secure connections during firefighting efforts.

Why are adapters important in firefighting equipment?

Adapters ensure compatibility between different hoses and appliances, facilitating seamless operation across various fire department equipment.

What role do strainers and jet siphons play in firefighting?

Strainers prevent debris from entering pumps, while jet siphons assist in transferring water quickly between tanks, maximizing available resources.

What are some common valves used in firefighting and what is their function?

Valves from brands like Harrington, Akron, and Elkhart help control water flow direction and pressure during firefighting operations.

What is the role of cutting tools like power saws and hydraulic cutters in firefighting?

Cutting tools are used for extrication, ventilation, and creating access routes in rescue operations, such as cutting through walls or debris.

How do piercing nozzles help in fighting fires in confined spaces?

Piercing nozzles allow firefighters to penetrate walls or roofs to deliver water directly to concealed flames, especially in building collapses or attic fires.

Why is ongoing training important for using cutting tools in rescue operations?

Proper training ensures that firefighters use cutting tools safely and efficiently, minimizing risks and ensuring timely access to victims during emergencies.

Chapter 6: Fire Hose Handling Techniques

Handling a fire hose is a vital skill for any firefighter, demanding both precision and strength. At its core, this task involves maneuvering powerful streams of water to quell dangerous flames, an endeavor that requires not only physical prowess but also meticulous technique. When confronted with the formidable force of a charged line, firefighters must employ their bodies expertly, relying on their core and legs to absorb impact and maintain control. Beyond mere brawn, the art of handling fire hoses encompasses a symphony of strategy and coordination, turning what might seem as mundane as holding a hose into an exercise in professional agility.

In this chapter, we dive into the essential techniques needed to master fire hose handling. Readers will explore the foundations of basic grip methods, learning how to distribute weight effectively to minimize fatigue during operations. The text delves into teamwork dynamics, illustrating how precise communication is crucial in deploying hoses swiftly and safely amidst the chaos of a fire scene. Safety considerations are also paramount, with guidelines on avoiding common hazards associated with hose management and ensuring personal protective equipment is utilized efficiently. Equipped with these insights, firefighters can not only hone their skills but also enhance their overall safety and effectiveness in the field.

Introduction to Fire Hose Handling

FIRE HOSE HANDLING is an indispensable skill for firefighters, as it plays a critical role in effectively suppressing fires. Mastery of this technique not only ensures that fire suppression efforts are successful but also significantly enhances the safety and efficiency of operations under high-stress conditions. By focusing on three core aspects – basic handling, teamwork, and safety – firefighters can develop the necessary skills to manage hoses adeptly and respond efficiently to various fire scenarios.

Understanding the importance of mastering fire hose handling is fundamental. In firefighting, time is of the essence, and proper hose management can make the difference between a contained blaze and an uncontrolled inferno. A firefighter who has mastered these techniques can swiftly maneuver the hose to target flames directly, maintaining control over water flow and pressure. This level of proficiency requires dedication and practice, allowing the firefighter to operate intuitively even when the environment is chaotic.

Basic handling techniques form the bedrock of effective fire hose use. It begins with grip methods that distribute the hose's weight evenly across the body, minimizing fatigue and maximizing control. Firefighters are taught to engage their core and leg muscles primarily, leveraging the body's largest muscles to handle the intense reaction forces generated by flowing water. Positioning the legs in an 'athletic stance' creates a stable base, enabling efficient nozzle operation while reducing strain on the arms and shoulders (Brumley, 2019). Additionally, maintaining an appropriate length of hose in front of the operator enhances flexibility and reduces the effort needed to direct water streams effectively. Adjustments may be necessary depending on the type and size of the hose, ensuring optimal control as conditions change.

Teamwork is equally crucial in fire hose operations. Effective communication and coordination within the team facilitate smooth hose deployment and movement. Each member plays a vital role, from the nozzle operator to backup firefighters. The latter support the former by managing hose drag and aiding in advancing the line, which allows the nozzle operator to focus on delivering water onto the fire. As highlighted in Source 2, efficient hose deployment is integral to getting firefighters into position quickly, minimizing the time spent outside and maximizing the time spent firefighting where it matters most (Fire

Spotlight, 2022). Pre-planning and training on hose deployment ensure that firefighters can adapt to various layouts and obstacles encountered in different building structures, ultimately improving response times and effectiveness.

Safety considerations are paramount when handling fire hoses. Proper techniques serve not only to improve efficiency but also to prevent injuries. Firefighters must be aware of potential hazards such as kinks, twists, or excessive bends in the hose, which could impede water flow or create dangerous recoil forces. To mitigate these risks, regular training and practice sessions are essential, reinforcing safe practices and muscle memory. Furthermore, using personal protective equipment (PPE) appropriately is vital to safeguard against burns, cuts, and other injuries that might occur during hose operations. Understanding these safety protocols helps maintain operational effectiveness while minimizing health risks to firefighters.

An integrated approach to fire hose handling combines basic techniques, teamwork, and safety considerations to optimize performance and efficiency. By emphasizing these interconnected aspects, firefighters can enhance their ability to manage hoses under pressure, reduce response times, and increase overall success rates in fire suppression. This holistic method not only equips firefighters with practical skills but also fosters a culture of continuous improvement and adaptation in response to evolving firefighting challenges.

Basic Techniques in Hose Handling

NAVIGATING THE INTRICACIES of fire hose handling is a crucial skill for firefighters, especially when it comes to advancing both charged and uncharged lines. Understanding these fundamentals not only ensures effective firefighting efforts but also enhances safety on the field.

First, it's essential to differentiate between charged and uncharged hoses. A charged hose is filled with water, which significantly increases its weight and the pressure within it. This can make maneuvering more challenging compared to an uncharged hose, which is lighter and devoid of pressure. Knowing how to handle each type effectively is a key aspect of firefighting.

When dealing with an uncharged line, coordination is vital. The relative ease of handling allows for quicker positioning; however, the risk lies in potential tangling or kinking if not managed carefully. Team members must work harmoniously to ensure the hose is laid out without twists or loops that could impede the flow of water once charged. This often involves strategic placement and communication among the team to anticipate and correct any emerging issues before they escalate.

Conversely, advancing a charged line demands a different set of skills due to its significant weight and internal pressure. Stability and control become critical. Firefighters need to employ their strength and technique to counterbalance the hose's heft. One effective method involves using body weight to anchor the hose while maintaining a firm grip to prevent sudden shifts that could destabilize the operation or cause injury. The working line drag and other methods described can be invaluable for maintaining control (*EFF I Chapter 15*, 2020).

A fundamental guideline in managing both charged and uncharged hoses is preventing kinks and twists. A kinked hose can severely restrict water flow, compromising firefighting operations. Techniques for prevention include manual straightening where feasible and repositioning hose segments strategically to ensure smooth bends. More importantly, anticipating potential problem areas before charging the line can save time and effort during critical moments.

Firefighters use specific methods to handle hoses efficiently, tailored to the situation at hand. For instance, stairways present unique challenges. While advancing uncharged hoses up stairs can be straightforward due to their lighter weight, charged hoses require careful planning. Using gravity and teamwork to manage the additional weight can make this task safer and more efficient (*Stretching and Advancing the Initial Handline*, 2014).

Safety and Teamwork Considerations

IN FIRE HOSE OPERATIONS, various team roles are critical for ensuring smooth and effective firefighting efforts. The first role to discuss is that of the nozzle operator. This individual is at the frontline, tasked with managing water streams and adjusting the flow based on the fire's behavior. Steadiness under pressure is crucial here since any mismanagement can disrupt the suppression strategy. For example, if a nozzle operator loses control, it could lead to ineffective water distribution, allowing the fire to spread further. Thus, maintaining a firm grip and practicing controlled movements is essential.

Supporting the nozzle operator are backup firefighters. Their primary responsibility is to manage hose drag and facilitate the hose's advancement towards the fire. This might involve pulling and adjusting the hose to ensure it's free of kinks and tangles that can hinder water flow. In large-scale operations, where hoses need to cover long distances, the role of backup firefighters becomes increasingly vital as they help manage the added weight and friction. Team coordination is key; each member must clearly understand their duties and work in harmony to maintain a continuous and effective attack line.

Additionally, situational awareness among team members plays a pivotal role during extensive firefighting operations. As conditions evolve, such as shifts in wind direction or structural changes, team members must adjust their positions accordingly. This adaptability not only ensures the safety of personnel but also maximizes the efficiency of the operation. For instance, in a scenario where the fire rapidly spreads due to a gust of wind, team members should be prepared to re-evaluate their positioning and tactics swiftly.

Safety during hose handling involves both preventing injuries and ensuring proper use of personal protective equipment (PPE). Common injuries include strains from lifting heavy hoses and slips due to wet surfaces. It's important to adopt techniques such as using legs rather than back muscles for lifting and implementing non-slip footwear to mitigate these risks. According to the National Fire Protection Association, improper or absent PPE significantly contributes to firefighter injuries (Firefighter Personal Protective Equipment (PPE), 2022). Therefore, adherence to protective gear protocols is non-negotiable.

When discussing protective gear, helmets, gloves, boots, and respiratory protection like SCBAs are fundamental. Helmets shield against falling debris, while gloves protect hands from burns and cuts. Boots equipped with steel toes guard against heavy objects, and SCBAs offer essential protection in smoke-filled environments. Proper maintenance and readiness of this gear are as crucial as their usage. Regular checks for wear and tear ensure reliability when equipment is most needed.

During retreats, safety considerations become even more pressing. When a situation demands withdrawal, whether due to escalating danger or strategic repositioning, it's vital that all team members understand the procedures involved. Retreats must be executed seamlessly to avoid confusion or injury. A sound guideline here includes clear communication from leadership, ensuring everyone is aware of exit routes and regrouping points. This organized approach reduces panic, helping personnel exit safely and regroup efficiently.

Bringing It All Together

IN THIS CHAPTER, WE have delved into the essential skills required for firefighters to handle fire hoses safely and efficiently. Understanding the core elements of basic handling techniques lays a strong foundation for effective hose management. By mastering grip methods and engaging key muscle groups, firefighters can control reaction forces and minimize strain during operations. The importance of teamwork has also been emphasized, illustrating how coordinated efforts enable smooth hose deployment and reduce risks associated with tangles or kinks. Coupled with safety protocols, such as adhering to PPE guidelines and regular training, these practices ensure the well-being of firefighters while enhancing their performance during critical moments.

With a clear focus on both individual skills and collective responsibility, firefighters can approach fire suppression with confidence and proficiency. Building upon the principles discussed, adaptability becomes a vital asset, allowing teams to adjust strategies in dynamic environments. As firefighters continue to face evolving challenges, ongoing practice and commitment to improving hose-handling techniques will remain crucial. By fostering a culture of learning and preparedness, the firefighting community can strengthen its ability to protect lives and property effectively, ensuring that every operation is a testament to skillful execution and unwavering dedication.

Chapter 6 Questions and Answers

Why is mastering fire hose handling crucial for firefighters?

Mastering fire hose handling is crucial because it directly impacts the efficiency and safety of fire suppression efforts, ensuring that firefighters can respond effectively and safely in high-stress situations.

What is the importance of grip methods in fire hose handling?

Grip methods are vital to evenly distribute the hose's weight, minimizing fatigue and maximizing control over the hose during firefighting operations.

What role does body positioning play in fire hose handling?

Proper body positioning, such as maintaining an 'athletic stance,' helps create a stable base, reducing strain on the arms and shoulders while enhancing the firefighter's ability to operate the nozzle.

How can firefighters prevent fatigue when handling fire hoses?

Firefighters can prevent fatigue by engaging their core and leg muscles, using their largest muscle groups to manage the hose's reaction forces instead of relying on their arms and shoulders.

What is the difference between a charged and uncharged hose?

A charged hose is filled with water, which increases its weight and pressure, making it more challenging to handle, whereas an uncharged hose is lighter and easier to maneuver.

What are the challenges when advancing a charged hose?

The main challenges with advancing a charged hose include its increased weight and pressure, which require more strength and coordination to maintain control while moving the hose towards the fire.

Why is teamwork essential in fire hose handling?

Teamwork is crucial for smooth hose deployment and advancement. Each firefighter plays a role in managing hose drag, ensuring the nozzle operator can focus on controlling the water stream.

How do backup firefighters assist during hose operations?

Backup firefighters manage hose drag and help advance the line, ensuring the hose remains free of kinks and tangles, which could impede water flow.

What is the importance of situational awareness in fire hose operations?

Situational awareness allows firefighters to adapt to changing conditions, such as wind shifts or structural changes, which can affect firefighting tactics and positioning.

What safety considerations should firefighters keep in mind during hose handling?

Firefighters must avoid common hazards like kinks, twists, or excessive bends in the hose, which can disrupt water flow or create recoil forces. Proper PPE should also be worn to minimize injury risks.

What role does personal protective equipment (PPE) play in fire hose handling?

PPE, including helmets, gloves, boots, and SCBAs, protects firefighters from injuries like burns, cuts, and falls while handling hoses in hazardous conditions.

What injuries can occur during fire hose handling, and how can they be prevented?

Injuries such as strains from lifting heavy hoses and slips due to wet surfaces can occur. These can be prevented by using proper lifting techniques, wearing non-slip footwear, and adhering to PPE protocols.

Why is it important to avoid kinks and twists in the fire hose?

Kinks and twists restrict water flow, compromising firefighting efforts. Preventing them ensures consistent water pressure and stream control, which is vital for effective fire suppression.

How does body weight help when advancing a charged hose up stairs?

By using body weight to anchor the hose, firefighters can reduce the amount of force needed to control the hose, making the task safer and more efficient, especially when working with a charged line.

What are the benefits of pre-planning hose deployment in fire scenarios?

Pre-planning hose deployment ensures that firefighters can quickly adapt to various building layouts and obstacles, reducing response time and increasing efficiency during fire suppression.

What is the role of the nozzle operator in fire hose operations?

The nozzle operator is responsible for managing the water stream, adjusting the flow based on the fire's behavior. Maintaining a steady grip and controlled movements is crucial to effective fire suppression.

What should be done when a situation requires a retreat from the fire scene?

When retreating, clear communication and awareness of exit routes and regrouping points are essential. This ensures a safe and efficient withdrawal from the fire scene.

How do firefighters prevent slips during hose handling operations?

Firefighters can prevent slips by wearing non-slip footwear and being mindful of wet and slippery surfaces during operations.

How does effective communication contribute to safe hose handling?

Effective communication ensures that each team member understands their role and can adjust actions in response to changes in the fire situation, ensuring smooth hose deployment and safety.

Why is regular training important for fire hose handling techniques?

Regular training helps reinforce safe practices, improve muscle memory, and ensure that firefighters are prepared to handle hoses efficiently under pressure, reducing the risk of injury and improving operational effectiveness.

Chapter 7: Types of Hose Deployments

Deploying hoses is at the heart of efficient firefighting techniques. The ability to swiftly and accurately manage hose deployment can make all the difference in controlling a fire situation effectively. In this chapter, we delve into various hose deployment strategies that enhance firefighting operations, focusing on both speed and precision. From navigating tight corridors to tackling intense blazes, the tactical use of hoses plays an essential role in ensuring firefighter safety and maximizing the impact of suppression efforts. Every method carries distinct advantages, tailored to meet the challenges encountered on diverse firegrounds. Understanding these fundamental tactics is crucial for any team looking to improve their operational efficiency in high-pressure scenarios.

This chapter will explore several critical approaches to hose deployment, offering insights into standard methods like offensive attack lines that take direct action against fire cores within structures. We'll also look at defensive water curtains, which serve as essential barriers preventing the spread of flames to surrounding areas. Furthermore, specialized strategies such as transitional and blitz attacks will be discussed, demonstrating how they adapt to changing conditions and harness large volumes of water for rapid extinguishment. Lastly, we'll compare the use of pre-connected versus static hose lines, highlighting their respective benefits and limitations based on scenario-specific needs. Each section aims to provide valuable knowledge and practical guidance for firefighters seeking to refine their craft and elevate their response capabilities in the face of unpredictable challenges.

Standard Deployment Methods

IN THE REALM OF FIREFIGHTING, effective hose deployment is crucial for efficient fire suppression and protection operations. Understanding standard methods can make the difference between quick control and prolonged engagement with a fire. One of the primary strategies involves using offensive attack lines, which are designed to deliver water directly to the heart of the fire.

Offensive attack lines are integral to swiftly suppressing the main body of a fire. The goal here is speed and efficiency, targeting the fire's core to minimize damage and prevent further escalation. This method relies heavily on advancing hoses into structures—a technique that requires not just speed but precision. When firefighters move these hoses inside, the focus is on reaching the seat of the fire, where flames are most intense and uncontrolled. By directing water here, firefighters can quickly lower temperatures and halt the spread, effectively gaining an upper hand in the operation.

The success of advancing hoses into structures hinges on several factors: the layout of the building, the fire's intensity, and the strategic placement of personnel. It demands close coordination among team members to ensure hoses are maneuvered through doorways or narrow passages without delay. Moreover, maintaining a steady supply of water to these hoses is vital; any interruption can compromise the operation and endanger both the firefighters and the structure's integrity.

Beyond offensive strategies, defensive methods play a critical role in firefighting, particularly through defensive water curtains. These serve as protective measures rather than direct extinguishing tools. Imagine a scenario where a fire threatens to leap across boundaries and ignite adjacent structures. Here, defensive water curtains become invaluable. By creating a barrier of finely atomized water droplets, they absorb the heat radiating from the fire, significantly reducing the risk of ignition nearby.

Water curtains are particularly advantageous in specific contexts, such as wildland-urban interfaces and industrial settings, where the risk of rapid fire spread is heightened. In wildland-urban areas, nature meets civilization, presenting unique challenges when fires break out. A water curtain can act as a shield, safeguarding residential properties while controlling the fire's expansion. Similarly, in industrial environments where flammable materials abound, deploying water curtains can help contain the blaze within set boundaries, protecting critical infrastructure and assets.

The creation of defensive water curtains involves precise calculation and execution. Firefighters must assess environmental factors like wind direction and intensity to position the curtains effectively. The equipment used—hoses with specialized nozzles that disperse water in fine mists—must be deployed with care, ensuring adequate coverage without depleting the water supply prematurely.

Guidelines for deploying attack lines in interior firefighting are essential to understand. Firstly, preparation is key. Firefighters need to ensure their hoses are devoid of kinks and fully charged before entering a structure. Secondly, communication is vital throughout the operation. Team members should maintain visual contact whenever possible or use radios to coordinate movements and actions. Thirdly, awareness of surroundings cannot be overstressed. Knowing exit routes and being vigilant about potential hazards like collapsing debris or sudden flare-ups are critical to safety.

Deploying hoses effectively also means understanding the dynamics of the fire. For example, directing water at the base of the flames, rather than the tops, ensures that cooling occurs where it's needed most. Additionally, using straight streams or narrow fog patterns can penetrate deeper into the burning material, facilitating quicker extinguishment.

Both offensive and defensive hose deployments require ongoing assessments and adaptations by firefighting teams. As conditions evolve, so too must their strategies. Whether advancing hoses into the inferno or setting up water curtains to block the heat, each action must be deliberate and informed by the circumstances at hand.

Specialized Deployments

TRANSITIONAL ATTACKS and blitz attacks are specialized strategies designed to address varying challenges on the fireground, each offering distinct advantages in firefighting operations. These tactics have been developed to enhance safety and effectiveness, optimizing the deployment of fire hoses during intense situations.

Transitional attacks initiate with exterior water application, targeting the fire's heat and intensity from outside the structure before advancing inside for further suppression efforts. This method is particularly effective when dealing with aggressive fires that threaten both property and firefighter safety. By cooling the environment from an external position, transitional attacks can significantly reduce conditions conducive to flashover, a dangerous event where everything in a room ignites almost simultaneously due to extreme heat. The initial exterior application helps manage the thermal balance within the structure, allowing crews a safer entry point to conduct interior operations. This approach not only improves visibility by decreasing smoke levels but also mitigates the risk of structural collapse, providing firefighters a strategic advantage in maintaining control over the fireground.

Once the initial attack reduces the fire's intensity, teams transition into the building to target remaining hotspots and fully extinguish the fire. This method serves dual purposes: it enhances firefighter safety and maximizes efficiency in fire containment. For instance, in residential fires, applying water from a safe distance initially can cool down rooms sufficiently, allowing firefighters to enter and perform necessary rescue operations or handle interior flames with reduced exposure to heat and toxic gases. It's crucial that teams on the ground maintain clear communication and coordination to ensure the smooth transference from exterior to interior operations, emphasizing the importance of practice and preparedness. Mastery of this tactic demands frequent training exercises to simulate various scenarios, ensuring readiness for real-world applications (Gettemeier, 2023).

Blitz attacks complement transitional strategies by employing high-flow nozzles to deliver large volumes of water rapidly. This tactic aims to overwhelm the fire quickly, making it especially effective against severe blazes that could otherwise escalate uncontrollably. Blitz attacks leverage the power of master streams, sometimes deploying impressive amounts of water in short durations to achieve a swift knockdown of the fire. This rapid response is critical in urban and industrial settings, where delays can lead to catastrophic losses due to high property values and dense populations.

In implementing blitz tactics, equipment such as portable monitors becomes indispensable. These devices allow for the quick setup and operation of powerful streams, adaptable to various angles and distances, thus providing coverage where handlines might be less effective. Portable monitors offer versatility; they can be operated remotely or manually adjusted to focus water precisely where it is needed most, whether that be through windows, doors, or directly onto exterior flames. This equipment's ability to sustain high-flow rates without physical strain on personnel means that firefighting teams can manage their resources more effectively, allowing them to focus manpower on other essential tasks such as search and rescue or securing the perimeter.

Fire departments must train regularly on these strategies, including the integration of portable monitors into their standard operating procedures. Training should involve hands-on sessions with equipment, simulations of urban and industrial setups, and learning to adjust to the dynamic nature of fire spread under varying weather conditions (Tactics for Deck Gun Blitz Attacks, 2024). Moreover, understanding the limitations of blitz attacks is equally important. While they can deliver impressive results, teams must ensure adequate water supply and pressure to sustain the deluge required for successful suppression. Coordination between water source providers and tactical units is imperative for sustained operations, highlighting the intersection of logistics and on-the-ground execution.

Both transitional and blitz attacks underscore the necessity of adaptive strategies in fireground operations. As every fire presents unique challenges, the flexibility and precision of these deployments make them invaluable tools in a firefighter's arsenal. By understanding and implementing these techniques effectively, fire departments can enhance their operational success, safeguarding lives and property more efficiently. Furthermore, the implications of these tactics extend beyond immediate firefighting needs, contributing to broader strategic planning for future incidents. The lessons learned from each deployment inform improvements in equipment, training, and inter-agency cooperation.

Pre-connected vs. Static Lines

IN FIREFIGHTING OPERATIONS, the choice between pre-connected and static hose lines plays a crucial role in determining how effectively teams can respond to various fireground situations. Understanding their operational efficiency and adaptability can significantly impact the outcomes of fire suppression efforts.

Pre-connected hose lines are renowned for their efficiency in quick deployment, offering rapid responses during smaller incidents. They typically come in fixed lengths that are pre-mounted on fire apparatus, making it easy for firefighters to swiftly unroll and engage with minimal setup time. This quick action capability is especially valuable in fast-evolving scenarios, such as when handling fires in tenements, garden apartments, or private dwellings (Staff, 2000). The immediate readiness of pre-connects allows firefighters to initiate an aggressive interior attack promptly, often serving as the go-to solution when engine companies need to react quickly in district fires.

However, the fixed length of pre-connected lines can also be a limitation. They may not suffice for larger or more complex firegrounds where the distance from the water source to the fire point exceeds the preset hose length. In these instances, firefighters must reassess their strategy to avoid problems like excessive kinking or reduced water flow due to overestimating or underestimating hose stretches (Staff, 2000).

On the other hand, static hose lines provide flexibility in terms of length, thereby offering greater adaptability in diverse and challenging scenarios. Unlike pre-connected lines, static hoses are not restricted by length; they allow firefighters to tailor the amount needed based on specific tactical needs and environmental conditions. This feature makes them particularly advantageous in larger operations, such as those involving extensive buildings or complicated layouts.

While static lines undoubtedly bring benefits in adjustability, they require more time and effort to set up compared to their pre-connected counterparts. Firefighting teams must carefully estimate the distance required and manually connect the appropriate number of hoses to ensure effective reach. This additional setup phase demands well-coordinated teamwork and communication. The nozzle firefighter needs to conduct a thorough size-up and stretch estimate upon arrival, deciding on the necessary working lengths to successfully tackle the fire at various floors or hidden spaces (Smith, 2013).

One example of this flexibility is seen when static lines are deployed off the rear step of the engine to efficiently navigate around structural obstacles or longer distances to the fire scene. This approach prevents overstretching, reducing potential challenges like friction loss that could impede water delivery to the desired location (Smith, 2013).

Ultimately, choosing between pre-connected and static hose configurations involves considering several critical factors. Fireground complexity is a primary consideration—whether the incident occurs in an expansive factory building, a high-rise structure, or a confined residential space dictates the suitable hose line type. Additionally, available resources, including manpower and equipment, influence decision-making. A static line might be more feasible in scenarios where ample personnel can support the extended setup time and coordination needed for longer stretches.

Strategic objectives also guide configuration choices. If speed and immediacy are paramount, a pre-connect might be preferred despite its length restrictions. In contrast, when precise control and maneuverability are essential, such as in cases requiring navigation through unpredictable environments or obstacles, static lines offer significant advantages.

Bringing It All Together

IN THIS CHAPTER, WE explored the essential strategies for deploying fire hoses efficiently, focusing on both offensive and defensive tactics crucial in firefighting operations. We discussed the importance of offensive attack lines to quickly target the core of a fire, enabling swift suppression and reducing potential damage. Equally important are defensive measures like water curtains, which protect nearby structures from the heat and prevent fire spread in challenging environments, such as wildland-urban areas and industrial settings. Through understanding standard methods and their application, firefighters can enhance operational effectiveness and safety.

We also examined specialized deployment techniques like transitional and blitz attacks, which provide additional strategies for handling different fireground scenarios. Transitional attacks balance exterior and interior efforts, ensuring safer entry with reduced risk of structural collapse, while blitz attacks use high-flow nozzles to quickly overwhelm intense fires. The choice between pre-connected and static hose lines further demonstrates the need for adaptability, with each type offering unique benefits suited to specific conditions. By integrating these techniques into their skill set, firefighters can better address the diverse challenges they face, optimizing fire suppression and protection efforts.

Chapter 7 Questions and Answers

What is the primary goal of offensive attack lines in firefighting?

The primary goal is to deliver water directly to the core of the fire to suppress it quickly, minimizing damage and preventing further escalation.

How does effective hose deployment impact firefighting efficiency?

It ensures faster suppression of the fire, improving firefighter safety, reducing property damage, and maintaining control over the fire situation.

What is the key advantage of using defensive water curtains?

Defensive water curtains create a barrier to absorb heat from a fire, preventing flames from spreading to adjacent structures and protecting the surrounding area.

Where are defensive water curtains particularly useful?

They are especially effective in wildland-urban interfaces and industrial settings where rapid fire spread is a significant concern.

Why is the layout of a building important when deploying offensive attack lines?

The layout determines the ease of accessing the fire's core, influencing the speed and safety of hose deployment within the structure.

What is the purpose of directing water at the base of flames during fire suppression?

Directing water at the base ensures cooling where the fire's heat is most concentrated, effectively reducing the fire's intensity and stopping its spread.

What is a transitional attack in firefighting?

A transitional attack involves applying water externally to reduce the fire's heat before entering the structure to continue interior firefighting efforts.

How do transitional attacks improve firefighter safety?

By cooling the structure externally, transitional attacks reduce the risk of flashover and structural collapse, providing a safer environment for interior operations.

What equipment is typically used during a blitz attack?

High-flow nozzles and portable monitors are used to deliver large volumes of water rapidly, overwhelming the fire quickly.

What is the key difference between pre-connected and static hose lines?

Pre-connected lines are pre-mounted on fire apparatus for rapid deployment, while static lines offer flexibility in length but require more setup time.

When should a pre-connected hose line be used?

Pre-connected lines are ideal for smaller, fast-evolving incidents where quick response is needed, such as fires in residential areas.

What is a limitation of pre-connected hose lines?

Their fixed length may not be sufficient for larger or more complex firegrounds, requiring adjustments or additional lines.

Why are static hose lines more flexible than pre-connected lines?

Static lines can be extended or shortened based on the fire's specific needs, offering more flexibility in large or complex environments.

What is the main challenge when setting up static hose lines?

They require more time and effort to set up, as firefighters must manually connect hoses and estimate the necessary length.

In what scenarios are static lines particularly beneficial?

Static lines are ideal for large operations, such as fires in high-rise buildings or industrial settings, where flexibility in hose length is crucial.

What is a blitz attack's primary benefit?

It provides a rapid and overwhelming suppression of severe fires by using large volumes of water in a short amount of time.

How do portable monitors assist in blitz attacks?

Portable monitors allow firefighters to deliver high-flow streams of water from a distance, covering a larger area with minimal effort.

What factors influence the choice between pre-connected and static hose lines?

Factors include the complexity of the fireground, available resources, and the need for speed versus precision in deploying hoses.

How do defensive water curtains differ from direct firefighting efforts?

Defensive water curtains are used to protect areas from the heat of a fire, while direct firefighting efforts focus on extinguishing the fire itself.

What training is required to effectively implement transitional and blitz attacks?

Firefighters need regular practice with equipment, simulations, and scenario-based exercises to master the safe and effective use of these tactics.

Chapter 8: Types of Hose Loads for Deployment

Organizing and deploying fire hoses effectively is essential in firefighting operations. The way hoses are loaded impacts the speed and efficiency with which firefighters can respond to emergencies. Different configurations of hose loads each serve particular purposes, providing unique advantages as well as posing distinct challenges. Understanding these variations is a key part of optimizing overall firefighting strategy. This chapter will explore the nuances of various hose load techniques, demonstrating how they contribute to successful firefighting operations.

In this chapter, readers will be guided through an examination of common hose load types used by fire departments across different scenarios. Starting with the foundational Flat Load, you'll see why its straightforward design remains popular and effective for rapid deployment and easy repacking. The chapter also delves into more compact options like the Accordion Load and Horseshoe Load, analyzing their space-saving benefits and potential complications if deployed improperly. Finally, specialized pre-connect loads such as the Minuteman and Triple-Layer Loads are discussed, highlighting their efficiency in swift, tactical deployments. By focusing on the characteristics, applications, benefits, and restrictions of each configuration, firefighters gain valuable insights into selecting and executing the appropriate load type for diverse operational needs. This knowledge not only enhances immediate response capabilities but also fosters long-term proficiency and adaptability in emergency situations.

Common Hose Loads

IN FIREFIGHTING OPERATIONS, the method of organizing and deploying hoses is as important as having a highly trained team.

Different configurations offer varied benefits and challenges. As we delve into commonly used hose load configurations, understanding each type's nuances can greatly enhance operational efficiency on the fireground.

The Flat Load is perhaps the most prevalent setup among fire departments. This configuration involves laying the hose flat in layers within the bed of the fire apparatus. Firefighters appreciate the simplicity of the flat load for several reasons. First, it allows for smooth deployment, ensuring that hoses can be easily pulled and laid out quickly when arriving at a scene. Another advantage is its straightforward repacking process after use, which saves valuable time during cleanup and preparation for future deployments. However, it's crucial to note that while the flat load is suitable for long supply lines, particularly with large-diameter hoses, it may not be the best choice for compact storage solutions. This trade-off between ease of deployment and storage space becomes apparent during operations requiring large amounts of hose or when space within the apparatus is limited.

The Accordion Load presents a different strategy by folding the hose back and forth vertically. This creates a compact arrangement that conserves space within the apparatus while still allowing for quick loading. The accordion load's design means that hoses are stacked neatly in vertical folds, making it easier to handle and transport. However, this configuration demands careful attention during deployment to prevent kinking, which can hinder water flow and create delays in firefighting efforts. The challenge lies in managing the care needed while maintaining speed, a balance that requires practice and familiarity with the equipment. For many departments, especially those dealing with diverse terrains or urban environments, mastering the accordion load can lead to improved efficiency in tight spaces where hose handling becomes critical. According to Source 1, one drawback of the accordion load is that it can create sharp bends in the hose, exposing it to potential damage (Staff, 2010). It also tends to pack tightly, complicating deployment. Departments with wide, short hose beds sometimes find this load optimal for storing significant lengths of hose used in supply or attack lines.

The Horseshoe Load takes a unique approach, forming a U-shaped loop with the hose as it wraps around the edges of the hose bed. This design minimizes sharp bends, thus reducing stress points and potential for damage during transit. Its stability during transport makes it an appealing option for many fire departments. Additionally, the horseshoe load facilitates smoother deployment when precision is key. The horseshoe load's main challenge lies in its requirement for exact packing conditions. Both improper alignment and inconsistent loops can result in tangling. A well-packed horseshoe load provides a stable and reliable solution, making it an excellent choice for situations needing consistent hose control. According to Source 2, when using threaded couplings, such loads should ensure proper orientation to prevent complications during deployment (*Chapter 15: Handling Hose (VIDEO QUIZ 2) Flashcards*, 2024).

Each configuration—flat, accordion, and horseshoe—demonstrates distinct advantages tailored to specific scenarios. As fire departments choose their hose load arrangements, considerations include the type of incidents they frequently respond to, the nature of the environment, and the resources available, including manpower and equipment. Practicing different hose load techniques ensures readiness for a variety of emergencies, helping teams deploy quickly and efficiently under pressure.

Pre-connect Hose Loads

IN THE WORLD OF FIREFIGHTING, efficiency and speed are paramount when it comes to deploying fire hoses. This is where understanding specialized pre-connect hose loads becomes crucial for firefighters. Two common configurations that highlight these qualities are the Minuteman Load and the Triple-Layer Load. Both methods have their unique characteristics and benefits, making them essential tools in a firefighter's arsenal.

The Minuteman Load is celebrated for its ability to allow rapid single-person deployment, providing an effective solution for short attack lines. Its design emphasizes ease and speed, minimizing tangles that can impede a quick response. The minuteman load involves stacking the hose on the firefighter's shoulder, with deployment starting from the top. This approach not only ensures that the hose unfurls smoothly but also helps control the length of hose being used, thus preventing unnecessary kinks and entanglements during critical moments. Such methodology is particularly beneficial when navigating through stairwells or tight spaces, as it allows for swift and controlled movement. The load is often used in conjunction with a flat load to help manage long hosebeds, demonstrating its versatility (Kirby et al., 2010-02-01).

To effectively implement the Minuteman Load, it's important to follow a clear set of steps. First, after loading the hose, ensure that the coupling is securely positioned on the shoulder. As you move towards the target area, steadily pull from the top stack, allowing the hose to flake off naturally. This technique minimizes tangling and maintains a steady flow throughout the deployment process. Continuing this process until you reach the desired length ensures optimal usage of the hose without causing undue strain or delay. This structured approach promotes consistency, enhancing the overall efficiency of operations.

Meanwhile, the Triple-Layer Load offers another dimension of efficiency with its space-saving design. Stacking the hose into three layers, this configuration maximizes the available space within the hosebed, making it suitable for situations where compact storage is required. The triple-layer load can be deployed by a single firefighter, making it ideal for crews operating with limited manpower. Nevertheless, it demands careful packing to avoid tangling when extending the hose. Proper training in packing techniques is essential to leveraging the full potential of this method, as any oversight could lead to complications during deployment (Kirby et al., 2010-02-01).

The Triple-Layer Load, while efficient in its design, presents unique challenges that must be addressed to ensure smooth operation. Due to its layered structure, ensuring that each section of hose is neatly stacked and aligned is critical. During deployment, the firefighter should maintain a gentle yet firm grip on each layer to prevent premature unraveling or snagging. Additionally, consistent practice in loading and unpacking this configuration is key to mastering the technique, allowing for quick adaption in various operational scenarios.

Both the Minuteman and Triple-Layer Loads exemplify the emphasis on speed and efficiency that is vital in firefighting contexts. Efficient deployment hinges not only on the choice of load but also on the technique used in packing and deploying the hoses. These methods underscore the importance of practicing and refining deployment skills to ensure seamless operations on the field. Firefighters need to be adept at swiftly selecting the appropriate load type depending on the situation, and executing deployment with precision. This adaptability is essential when responding to fires in diverse settings, from urban high-rises to rural dwellings (Choosing Hose Loads with Multiple Options for Your Rural Fire Apparatus - Fire Apparatus: Fire Trucks, Fire Engines, Emergency Vehicles, and Firefighting Equipment, 2020).

Moreover, understanding the intricacies of these specialized pre-connect hose loads can greatly influence the outcome of firefighting efforts. For instance, scenarios requiring swift navigation through buildings or tackling fires in confined spaces benefit significantly from the Minuteman Load's maneuverability. Conversely, areas necessitating extended coverage might find the Triple-Layer Load advantageous due to its compactness and ease of deployment over longer distances. Each configuration has distinct attributes that cater to different firefighting needs, thereby optimizing the use of resources and time.

Additionally, the significance of proper packing cannot be overstated. A well-packed hose guarantees a hassle-free deployment, enabling firefighters to focus on the task at hand rather than dealing with entanglements or delays. Regular drills and intensive training on hose packing and deployment techniques are invaluable tools for maintaining proficiency and readiness. Fire departments must invest time and resources in ensuring that their teams are acquainted with multiple hose load configurations and confident in applying them during emergencies (Kirby et al., 2010-02-01).

Specialty Hose Loads and Influencing Factors

IN THE DYNAMIC WORLD of firefighting, the deployment of fire hoses is both an art and a science. Specialty hose loads are particularly critical in ensuring the success of firefighting operations. These loads must be chosen based on specific operational needs, terrain, and team coordination capabilities. This examination begins with the Reverse Lay technique, which is quintessential for rural or extensive firefighting operations.

The Reverse Lay method stands out as an efficient strategy for establishing a water supply line quickly, especially in rural areas where hydrants may be scarce or non-existent. Its primary advantage lies in its ability to set up a rapid water supply to the fire scene. In practice, this involves laying the hose from the fire scene back to the water source, usually a tanker or portable tank. The effectiveness of this method relies heavily on precise coordination among crew members, as well as seamless communication with pump operators who manage water flow and pressure. Considerations must also include the layout of rural roads or properties, which can influence the speed and efficiency of this deployment. Despite its demanding nature, when executed correctly, the Reverse Lay provides a vital lifeline in rural firefighting efforts, streamlining the logistics required to deliver water swiftly to remote locations. This method's success is contingent upon rigorous training and synchronization within the team, emphasizing the need for drills that replicate real-world scenarios.

Meanwhile, the Skid Load offers unique advantages tailored for wildland firefighting, a domain where agility and portability are paramount. This load configuration is designed for lightweight, portable deployment and excels in rugged terrains where traditional fire apparatus might struggle. Skid Loads typically involve smaller-diameter hoses that are easier to handle across uneven ground. They are ideal for quick attacks and maneuverability in forested areas, where trails are narrow, and access is restricted. Firefighters engaging in wildland operations often face unpredictable conditions such as gusting winds and rapidly spreading flames, necessitating a hose load that can be deployed rapidly and retracted just as quickly. This adaptability makes the Skid Load a favored choice for fire crews tasked with protecting vast tracts of land.

Furthermore, understanding the impact of apparatus design, crew size, and operational needs is crucial when selecting the appropriate hose load. Each fire department's equipment configuration and human resources profoundly influence the decision-making process. For instance, a compact fire engine might require more straightforward hose load methods to allow for quick access and deployment. Conversely, larger crews can manage more complex hose loads that might offer enhanced water flow rates but require multiple hands to execute efficiently. Operational needs also dictate the choice of hose load; urban fires with high-rise buildings necessitate different strategies compared to wildfires sweeping through remote forests. Therefore, balancing the speed of deployment with ease of handling becomes a fundamental aspect of firefighting strategy. This balance ensures that firefighters have access to water as quickly and efficiently as possible, regardless of the scenario they encounter.

While comprehensive guidelines often assist firefighters in mastering these techniques, practical experience remains irreplaceable.

Each firefighting scenario presents distinct challenges, requiring flexibility and improvisation alongside established protocols. Consequently, ongoing training and adaptation are necessary to fine-tune the deployment of specialty hose loads in response to evolving firefighting environments.

Bringing It All Together

IN EXAMINING THE ORGANIZATION and deployment of fire hoses, this chapter has highlighted the pivotal role these configurations play in firefighting operations. By exploring various hose load techniques such as the Flat Load, Accordion Load, Horseshoe Load, Minuteman Load, Triple-Layer Load, Reverse Lay, and Skid Load, we have delved into their unique advantages, challenges, and applications. These methods are integral to ensuring efficiency and effectiveness on the fireground, allowing fire departments to adapt their tactics based on incident type and environmental conditions. From quick setups in urban environments to strategic deployments in rural or wildland areas, understanding and mastering these hose load techniques is essential for any firefighting team.

As we move forward, it's crucial to recognize the importance of continuous training and practice in refining these skills. Firefighters must be adept at selecting and deploying the appropriate hose load configuration swiftly and accurately, optimizing their response to emergencies. The adaptability of these strategies ensures that firefighters are well-prepared to face diverse challenges, whether navigating tight spaces within buildings or tackling fires across expansive terrains. By investing time in drills and familiarization with different hose loads, fire departments can enhance their operational readiness and resilience, ultimately safeguarding both lives and property more effectively.

What is the primary benefit of the Flat Load?

The primary benefit of the Flat Load is its simplicity in deployment and repacking, allowing for quick and easy hose deployment and a straightforward cleanup process.

What is a key disadvantage of the Flat Load?

A key disadvantage of the Flat Load is that it may not be the best choice for compact storage solutions, as it can take up a lot of space within the apparatus.

How does the Accordion Load save space?

The Accordion Load saves space by folding the hose back and forth vertically, creating a compact, space-efficient arrangement that is easy to transport.

What is one significant challenge when deploying the Accordion Load?

One significant challenge when deploying the Accordion Load is preventing the hose from kinking, which can hinder water flow and delay firefighting efforts.

What does the Horseshoe Load aim to minimize?

The Horseshoe Load aims to minimize sharp bends in the hose, reducing stress points and the potential for hose damage during transport.

What is a potential issue with the Horseshoe Load?

A potential issue with the Horseshoe Load is that improper packing or inconsistent loops can lead to tangling during deployment.

What is the Minuteman Load designed for?

The Minuteman Load is designed for rapid single-person deployment of short attack lines, making it ideal for quick response in confined spaces or situations requiring swift movement.

How does the Minuteman Load help prevent tangling?

The Minuteman Load prevents tangling by having the hose stacked on the firefighter's shoulder, with deployment starting from the top to ensure smooth and controlled unwinding.

What is the Triple-Layer Load's primary advantage?

The primary advantage of the Triple-Layer Load is its space-saving design, allowing it to be stored compactly while still being deployable by a single firefighter.

What challenge does the Triple-Layer Load pose?

The challenge with the Triple-Layer Load is ensuring that each section of the hose is neatly stacked and aligned, as improper packing can lead to tangling during deployment.

What is the Reverse Lay technique used for?

The Reverse Lay technique is used to establish a water supply line quickly, particularly in rural areas where hydrants may be scarce or non-existent, by laying hose from the fire scene back to the water source.

Why is the Reverse Lay technique effective for rural firefighting?

The Reverse Lay is effective for rural firefighting because it enables a quick setup of a water supply, which is crucial in areas where water sources are far away or limited.

What type of firefighting is the Skid Load best suited for?

The Skid Load is best suited for wildland firefighting, where portability and agility are essential for navigating rugged terrains.

What makes the Skid Load particularly useful in wildland firefighting?

The Skid Load is useful in wildland firefighting because it involves smaller-diameter hoses that are easier to handle across uneven ground and are ideal for rapid deployment and retraction in forested areas.

How does apparatus design influence hose load selection?

Apparatus design influences hose load selection by determining the available space and configuration of the fire engine, affecting the choice of hose load that allows for quick access and deployment.

Why is coordination important when using the Reverse Lay method?

Coordination is crucial when using the Reverse Lay method because it requires precise teamwork between crew members and pump operators to ensure the water flow and pressure are properly managed during deployment.

How does the choice of hose load affect operational efficiency?

The choice of hose load affects operational efficiency by influencing how quickly the hose can be deployed, the amount of space available, and the ease of handling, all of which impact response times during emergencies.

Why is practice important for mastering hose load techniques?

Practice is important for mastering hose load techniques because it ensures firefighters can deploy hoses quickly and efficiently, avoiding tangling or delays during critical moments on the fireground.

How does the Minuteman Load benefit firefighting in confined spaces?

The Minuteman Load benefits firefighting in confined spaces by allowing a firefighter to maneuver easily while deploying the hose, especially in stairwells or tight areas, thanks to the smooth unrolling from the top stack.

What is the primary benefit of selecting the right hose load configuration?

The primary benefit of selecting the right hose load configuration is optimizing speed and efficiency during deployment, which is crucial for effective firefighting in various environments and scenarios.

Chapter 9: Advanced Fire Suppression Strategies

Adapting to advanced fire suppression scenarios is a crucial aspect of modern firefighting. Firefighters face a myriad of challenges, especially with the increasing urbanization and industrial activities that demand more sophisticated response strategies. As the nature of fire incidents evolves, so must the tactics employed to combat them. Mastering complex suppression methods not only involves technical prowess but also an astute understanding of various factors that influence each unique fire scenario. From building design to weather conditions, these elements require quick assessment and informed decision-making. The ability to adapt and utilize specialized techniques effectively in the field can make a significant difference in safeguarding lives and minimizing property damage.

This chapter delves into the intricacies of advanced fire suppression, focusing on three pivotal components: attack strategies, water additives, and master stream operations. It seeks to equip firefighters with the knowledge needed to implement efficient attack strategies like blitz attacks and rapid water application. The role of water additives such as foams and wetting agents will be examined to highlight their utility in enhancing fire suppression efforts. Furthermore, the strategic deployment of master stream operations underscores their significance in managing large-scale fires. Through this exploration, the chapter aims to foster an adaptive mindset among firefighters, ensuring they are well-prepared to tackle any challenging situation with expertise and confidence.

Introduction to Advanced Suppression Strategies

IN THE EVER-EVOLVING field of firefighting, adapting to advanced fire suppression scenarios requires a deep understanding of specialized techniques and decision-making. The core objective of this subpoint is to underline the immense value that mastering complex suppression methods holds in various fire scenarios. As urbanization and industrial activities intensify, so do the challenges faced by firefighters. Consequently, there is an increasing need for developing and applying advanced tactics tailored to meet these challenges effectively.

Exploring the significance of mastering complex suppression methods begins with the comprehension of diverse fire scenarios. Each fire incident presents unique characteristics influenced by factors such as building design, materials involved, and weather conditions. Firefighters must equip themselves with the knowledge and skills required to swiftly assess these variables and implement the most effective response strategies. A comprehensive grasp of complex suppression methods enables firefighters to make informed decisions that optimize resource utilization while minimizing risk to life and property.

The chapter will focus on three key components: attack strategies, water additives, and master stream operations. Attack strategies serve as the blueprint for managing fire scenes efficiently. They encompass methods such as blitz attacks and rapid water application, designed to quickly overpower fires before they escalate (Mark, 2020). Blitz attacks, in particular, illustrate an aggressive approach where large amounts of water are delivered rapidly to the fire's seat, much like a focused assault on the fire's core. This tactic emphasizes the importance of swift action, reducing the potential for the fire to grow unchecked.

Water additives play a crucial role in enhancing fire suppression outcomes. These additives, such as foams and wetting agents, are designed to alter the properties of water, making it more effective in extinguishing flames. By improving water penetration in dense materials or forming a protective barrier over flammable liquids, additives enhance the firefighter's ability to control complex fire situations. For instance, Class A foams are utilized for structural fires, maximizing water efficiency, while Class B foams target flammable liquid fires, preventing re-ignition.

Master stream operations involve the deployment of high-volume appliances like deck guns and portable monitors. These tools are invaluable in large-scale fires where massive water discharge is necessary. Deck guns allow for precision water delivery from apparatus stationed at safe distances, whereas portable monitors provide flexibility in targeting multiple areas of involvement. Understanding when and how to employ these tools is vital for successful firefighting operations and requires practice and familiarity with each tool's capabilities.

Integrating sophisticated tactics into modern firefighting practices is paramount. This integration represents not only technological advancement but also a shift towards proactive and strategic firefighting. Modern firefighting demands an adaptive mindset, where firefighters continuously refine their techniques and incorporate new knowledge gained from research and experience. It is crucial for fire departments to prioritize training programs that emphasize advanced tactics and ensure readiness for any scenario.

Emphasizing the necessity of sophisticated tactics extends beyond technical execution; it also involves cultivating a culture of safety and efficiency. Advanced strategies contribute to overall fire ground safety by reducing exposure time and lowering the risk of structural collapse or hazardous material release. Efficient use of resources, including water supply and manpower, is optimized through these refined techniques, resulting in faster incident resolution and minimized environmental impact.

A brief overview of how these advanced strategies enhance fire ground safety and efficiency underscores their intrinsic value. By equipping firefighters with a repertoire of adaptable techniques, departments can improve their response times and effectiveness significantly. This approach not only saves lives but also conserves valuable community assets by reducing fire spread and damage.

To summarize, the goal of this subpoint is to highlight the purpose and key areas of discussion in advanced fire suppression techniques. Mastering complex suppression methods tailored to diverse fire scenarios is essential for contemporary firefighting. The chapter outlines a focus on attack strategies, water additives, and master stream operations, emphasizing the necessity of integrating sophisticated tactics into modern practices. This integration enhances overall fire ground safety and efficiency, setting the standard for future firefighting endeavors (*Fire Science Chapter 17 Fire Control Flashcards*, 2024).

Direct vs. Indirect Attack Methods

IN THE COMPLEX FIELD of firefighting, understanding the differences between direct and indirect attack methods is crucial for effective fire suppression. These techniques are designed to adapt to various scenarios by considering factors such as fire size, location, building construction, and available resources. By recognizing these distinctions, firefighters can make informed decisions that maximize safety and efficiency on the fireground.

Direct attack is a fundamental method where water is applied directly onto the fire itself. This approach aims to achieve immediate cooling effects, suppressing the flames at their source. Direct attacks are typically most effective in situations involving low-intensity fires, particularly those with flame lengths under four feet (*Colorado Firecamp, Wildland Fire Suppression Tactics Reference Guide*, n.d.). Applying water or other agents directly onto the burning material can quickly reduce temperatures, helping to control the spread of the fire. This tactic is especially useful when dealing with fires burning in light fuels or fuels with high moisture content, as it allows firefighters to work closely to the fire's edge, ensuring a swift response (Editor, 2021).

A major advantage of direct attack lies in firefighter safety. By being close to the fire's edge, crews can maintain a tactical advantage and have ready access to escape routes through the already burned areas—an area referred to as "keeping one foot in the black" (Colorado Firecamp, Wildland Fire Suppression Tactics Reference Guide, n.d.). This proximity can also facilitate quick adjustments and immediate actions when unexpected changes occur in fire behavior. Direct attack methods are often favored for room-and-contents fires, where the main goal is to limit fire damage within an enclosed space, allowing for minimal spread beyond the initial area of ignition.

In contrast, the indirect attack provides an alternative strategy suited to more challenging fire scenarios. This involves applying water or other methods indirectly onto surrounding surfaces rather than directly onto the fire. The aim here is to generate steam, which helps reduce heat in areas where the fire may not be directly accessible (Editor, 2021). Indirect attack leverages barriers, both natural and artificial, to create defensible spaces and manage larger, more intense fires effectively. This method is often deployed in wildland settings where a direct approach might expose firefighters to undue risk due to the intensity and speed at which the fire spreads.

The decision to employ either a direct or indirect attack involves several critical considerations. One primary factor is the fire's size and location. Large fires in open spaces, such as forests or fields, may necessitate an indirect approach due to the challenges of containment. On the other hand, smaller fires in confined spaces, like buildings, could benefit from a direct attack to prevent further structural damage. Building construction plays a significant role; for example, structures with combustible materials may require different strategies compared to those made of steel or concrete, which could contain the fire longer.

Available resources also influence the choice of strategy. Limited water supply might restrict the feasibility of a direct attack, while the presence of aerial support could enhance the effectiveness of an indirect approach. Understanding these nuances helps tailor the suppression efforts to match the specific demands of each incident, thus optimizing resource utilization while minimizing risks to personnel.

Situational examples illustrate why and when each method proves most effective. For instance, in basement fires, where accessibility is limited and conditions are hazardous, an indirect attack might involve applying tactics to cool and extinguish the fire from above or adjacent spaces without exposing firefighters to unnecessary danger. Conversely, in a room-and-contents fire scenario, a direct attack can swiftly knock down the flames, limiting fire extension and smoke production inside the structure.

Guidelines for choosing the right strategy highlight the complexities involved in firefighting. A strategic decision must weigh factors such as fire behavior predictions, environmental conditions, and potential hazards. Firefighters should evaluate these elements continuously, adjusting plans based on changing dynamics at the scene. Effective communication and coordination among team members are essential to executing either strategy efficiently.

Moreover, combining direct and indirect techniques can sometimes provide the best outcome, utilizing the strengths of both approaches. Mixing tactics allows for flexibility and can address multiple aspects of a fire simultaneously, such as attacking the main body of the fire directly while using indirect methods to protect vulnerable exposures and cut off potential paths of spread.

Water Additives and Master Stream Operations

IN ADVANCED FIRE SUPPRESSION, understanding the role of water additives and master stream operations is paramount. These strategies can significantly enhance firefighting efforts by improving efficiency and effectiveness in various challenging scenarios. Let's delve into foam types, wetting agents, the principles of master stream operations, and how these components can be strategically deployed to tackle large-scale or complex fires.

Foam Types: Foam plays a critical role in fire suppression, particularly when it comes to varied fire types that demand different extinguishing approaches. Class A and Class B foams serve distinct purposes based on the nature of the fire they are meant to combat. Class A foam, often utilized in structural fires, is designed to address fires involving ordinary combustibles such as wood, paper, and cloth. This type of foam helps penetrate the burning material by breaking down water's surface tension, which not only aids in rapid extinguishment but also reduces the possibility of rekindling (Bane, 2019). In contrast, Class B foam targets flammable liquid fires. It works by forming a vapor-suppressing film on the surface of the burning liquid, thereby extinguishing the fire by smothering its vapors. This differentiation underscores the importance of selecting the appropriate foam for a given fire scenario, optimizing both safety and resource use.

Wetting Agents: Wetting agents provide another layer of sophistication in managing deep-seated fires in dense materials. These agents, defined by NFPA standards, are known for their ability to reduce the surface tension of water, thus increasing its penetration and spreading abilities (Staff, 2008). When applied, wetting agents allow water to reach deeper into the burning material, ensuring more effective suppression especially in cases where the fire is embedded within dense materials, like heavy timber or piled tires. The enhanced spreading capability of wetting agents means that less water is needed overall, which is beneficial in situations where water resources are limited or accessibility is an issue.

Master Stream Operations: Moving beyond water and additives, master stream operations constitute a fundamental tactic in the domain of high-volume firefighting. This strategy employs powerful appliances such as deck guns and portable monitors, which are capable of delivering large volumes of water at significant pressure. These systems are particularly useful in tackling large-scale fires, such as those found in industrial complexes or during defensive firefighting operations when direct firefighting inside a structure may be too dangerous (Bane, 2019). Master streams provide a versatile option for controlling the perimeter of a fire or cooling down structures and exposures, preventing the spread of heat to adjacent areas.

Deployment Strategies: Effective deployment strategies are crucial when using water additives and master stream operations. This involves balancing the volume of water or foam with the constraints of supply limitations, especially in industrial settings or during extensive defensive maneuvers. For instance, while deploying Class A foam in a structural fire, it's important to maintain an adequate supply to ensure continuous application until the fire's core is fully extinguished. Conversely, in the case of an industrial fire involving flammable liquids, selecting a Class B foam tailored to the specific chemical properties of the liquid is critical for effective suppression (Staff, 2008).

Using master streams effectively requires careful consideration of the apparatus placement to maximize coverage and efficiency. For example, establishing a command post that allows clear visibility and control over deployed master streams can help achieve optimal results. In cases where water supply infrastructure may not accommodate high-volume usage, establishing relays or drafting from secondary sources may be necessary to sustain operations. Additionally, coordinating with other emergency services to manage peripheral activities such as evacuation and scene safety is vital.

Summary and Reflections

AS WE'VE EXPLORED IN this chapter, adapting advanced suppression techniques is crucial for firefighters facing ever-evolving challenges. By understanding and applying these specialized tactics, such as strategic attack methods, effective use of water additives, and proficiency in master stream operations, firefighters can meet the demands of various complex fire scenarios with confidence. This comprehensive approach equips them with the skills needed to make quick, informed decisions that prioritize both safety and efficiency. Amidst rapid urbanization and industrial growth, mastering these techniques ensures firefighters are well-prepared to manage incidents, safeguarding lives and property.

The integration of sophisticated strategies into modern firefighting practices represents a proactive shift towards enhanced fireground safety. The focus on refining skills through ongoing training and research empowers fire departments to remain at the forefront of their field. Ultimately, by fostering an adaptive mindset and cultivating a culture of safety, firefighters can improve response times while minimizing risks. Through these efforts, we set new standards for future endeavors, optimizing resource utilization and reducing the environmental impact of fires. As this journey unfolds, it becomes clear that the path forward lies in continuous learning and embracing innovation.

Chapter 9 Questions and Answers

What is the primary focus of advanced fire suppression strategies?

The primary focus is to equip firefighters with the knowledge and techniques needed to effectively combat complex fire scenarios, considering factors such as building design, weather conditions, and available resources.

Why are advanced fire suppression tactics increasingly important?

Due to growing urbanization and industrial activities, fire incidents have become more complex, demanding more sophisticated response strategies to ensure effective fire suppression.

What are the three key components of advanced fire suppression?

Attack strategies, water additives, and master stream operations.

What is a blitz attack, and when is it used?

A blitz attack is a tactic where large amounts of water are applied rapidly to the fire's seat, aiming to overpower the fire quickly before it can escalate. It is used in situations where swift action is critical to control the fire's growth.

What role do water additives play in fire suppression?

Water additives, such as foams and wetting agents, enhance the effectiveness of water by improving its penetration into materials and preventing re-ignition, thus aiding in faster and more efficient fire suppression.

What is the difference between Class A and Class B foams?

Class A foam is used for structural fires involving ordinary combustibles like wood and paper, while Class B foam is used for fires involving flammable liquids and helps smother vapors to prevent re-ignition.

How do wetting agents improve fire suppression efforts?

Wetting agents reduce the surface tension of water, allowing it to penetrate deeper into dense materials, making it more effective in controlling deep-seated fires, particularly when water is limited.

What is a master stream operation, and when is it used?

Master stream operations involve using high-volume appliances like deck guns and portable monitors to deliver large volumes of water at high pressure. They are used in large-scale fires or defensive firefighting situations.

What are the advantages of using master stream operations?

Master streams allow for precise water delivery from a safe distance, making them essential for controlling large fires or cooling exposures in high-risk areas.

What factors must be considered when deploying master streams effectively?

Factors such as water supply limitations, apparatus placement, and coordination with other emergency services must be considered to maximize coverage and efficiency.

What is the difference between direct and indirect attack methods?

Direct attack involves applying water directly onto the fire to suppress it at its source, while indirect attack applies water or agents to surrounding surfaces to generate steam and reduce heat, often used for larger or more intense fires.

When is a direct attack most effective? A direct attack is most effective in situations involving small to medium fires, such as room-and-contents fires, where the goal is to suppress the flames quickly and minimize damage.

In what situations is an indirect attack more appropriate?

Indirect attacks are more appropriate for large fires, especially in open spaces like wildland fires, where direct access to the fire is too dangerous or impractical.

How do building materials influence the choice of attack method?

Combustible materials require a different strategy than non-combustible ones. Fires in buildings made of materials like wood may benefit from a direct attack, while fires in steel or concrete structures may require more strategic methods, like indirect attacks.

What is meant by "keeping one foot in the black" in fire suppression?

It refers to the practice of ensuring an escape route by staying close to the fire's edge, which has already been burned, to provide a safe area for retreat if necessary.

How does the use of water additives reduce the environmental impact of firefighting?

By improving water efficiency and effectiveness, water additives reduce the overall amount of water used, minimizing water wastage and potential environmental damage from runoff.

Why is situational awareness crucial in selecting the right suppression strategy?

Fire behavior can change rapidly, and selecting the right strategy involves continuous evaluation of factors like fire size, location, and the resources available to maximize efficiency and safety.

What are the safety benefits of using advanced suppression strategies?

Advanced strategies like blitz attacks and master stream operations help reduce firefighter exposure to heat and hazards, ensuring better safety while controlling the fire.

How do foam types enhance the firefighter's ability to control complex fire scenarios?

The correct selection of foam, whether Class A for structural fires or Class B for liquid fires, improves the suppression process by addressing specific fire needs, reducing re-ignition, and preventing further spread.

How does modern firefighting practice incorporate technological advancements in suppression techniques?

Modern firefighting integrates sophisticated tools and techniques, such as high-volume master stream appliances and advanced water additives, to improve efficiency and safety on the fireground, enabling a more proactive approach to firefighting.

Chapter 10: Maintaining and Evaluating Fire Suppression Systems

Maintaining and evaluating fire suppression systems is a core responsibility for any fire department, demanding a dedicated focus on ensuring equipment readiness and operational efficiency. The myriad components involved in these systems, ranging from hoses and nozzles to advanced machinery like foam systems and portable pumps, require meticulous care and attention. Each piece must be kept in peak condition through regular inspections and maintenance, where even minor wear can lead to significant challenges during emergencies. This chapter serves as a guide to firefighters, providing them with essential practices and insights to uphold the integrity of their fire suppression systems effectively.

Throughout this chapter, readers will explore comprehensive strategies for maintaining critical firefighting equipment. These include detailed routines for inspecting hoses and nozzles to identify potential vulnerabilities, the importance of employing proper cleaning techniques to prevent damage, and effective storage methods that preserve equipment longevity. Additionally, the chapter delves into appliance and tool care, underscoring the necessity of preventative maintenance schedules and thorough record-keeping to anticipate and address issues before they arise. The discussion extends to handling routines that minimize equipment stress and promotes an organizational culture invested in continuous improvement. By emphasizing both the practical aspects and the strategic foresight required, this chapter equips firefighters with the knowledge needed to ensure safety and reliability when it matters most.

Hose and Nozzle Maintenance

WHEN IT COMES TO FIRE suppression systems, hoses and nozzles are lifelines that must be kept in peak condition for safety and functionality. Regular inspections are a cornerstone of maintaining these components. Firefighters should systematically check hoses for any visible signs of wear and tear, such as cracks or weak spots. This includes examining seals and the connection points of nozzles, where wear can lead to leaks or malfunction during critical operations. A detailed inspection routine can help identify small issues before they escalate into bigger problems, ensuring the readiness and reliability of the equipment when it's needed most. Citing expert advice, regular checks for vulnerabilities like kinks or abrasions can significantly enhance the lifespan and effectiveness of fire hoses (midwesthose, 2024).

Proper cleaning techniques are equally vital. Using mild detergents, hoses should be washed to remove any debris or residue that might affect their performance. It's essential to dry the hoses thoroughly to prevent mold formation, which can compromise the hose's material integrity. Implementing a post-use cleaning process is crucial; leaving residues can lead to clogging and deterioration over time. Considering the nature of firefighting environments, where dirt and soot are prevalent, consistent cleaning not only supports optimal performance but also helps extend the service life of hoses and nozzles. These cleaning practices follow guidelines to ensure thorough maintenance (jspence@hartindustries.net, 2024).

Storage methods play a significant role in preserving the longevity of hoses and nozzles. Proper coiling and stacking are important to avoid unnecessary stress on the hose materials. Storing these components in a cool, dry place, away from direct sunlight, extreme temperatures, and chemicals, can prevent premature aging and damage. For instance, keeping hoses loosely coiled on racks minimizes kinking and preserves the structure for longer use. Additionally, securing nozzles properly during storage and transit is crucial to protect them from physical damage that could impair their function. By maintaining an organized storage system, firefighters can reduce wear and tear, effectively prolonging the usability of their equipment.

Handling also matters greatly. While transporting hoses and nozzles, care should be taken to avoid dragging them across rough surfaces, which can lead to abrasions and weaken the materials. Awareness of handling and using protective covers when necessary can mitigate potential damage from harsh environments. Such preventive measures are simple yet effective strategies to prevent damage caused by external factors.

Engaging in routine checks and cleaning procedures is part and parcel of efficient maintenance culture among firefighting teams. Checking pressure levels and ensuring compatibility with pumps prevent overloading and structural failure, safeguarding against sudden bursts or leaks. Monitoring the equipment's working conditions regularly allows for adjustments, ensuring that all components operate within safe parameters. This proactive approach not only reduces replacement costs but also ensures that the entire fire suppression system remains reliable under operation pressures.

Firefighters should be educated about these maintenance protocols. Training sessions can be beneficial in teaching operators how to handle hoses correctly, demonstrating the importance of avoiding sharp bends and twists that can degrade hose quality. Encouraging responsible use among team members fosters a culture of accountability and expertise, reducing the incidence of misuse that could jeopardize operational efficiency.

Implementing a robust maintenance schedule tailored to specific needs is a step forward in harnessing the full potential of fire suppression systems. This involves logging inspection records, tracking repair histories, and noting replacements. By analyzing these records, patterns in wear and degradation can be identified, allowing for better resource planning and procurement.

Appliance and Tool Care

ESTABLISHING COMPREHENSIVE care routines for firefighting appliances and tools is critical to ensuring the readiness and effectiveness of fire suppression systems. These systems, including standpipes, deck guns, foam systems, and portable pumps, are essential in safeguarding life and property during emergencies. Therefore, a well-structured routine involves not only physical inspections but also preventative measures and meticulous record keeping.

The development of preventative maintenance schedules forms the backbone of effective fire suppression system management. Regular checkups are vital to detect potential issues before they escalate into significant problems, potentially compromising safety during a fire emergency. For systems like standpipes and deck guns, it's important to schedule frequent inspections that focus on verifying mechanical functionality and testing operational reliability. Foam systems require specific attention on their components' integrity and the quality of the foam concentrate itself. Portable pumps, given their mobility and often rugged use, should undergo more frequent testing to ensure their mechanisms remain in optimal condition.

Guidelines for preventative maintenance should include visual checks for damage, verification of pressure levels, and function tests of moving parts. Establishing a routine, perhaps quarterly or semi-annually depending on usage intensity, can help identify any irregularities early on. By involving multiple personnel in these checks, you ensure greater oversight and reduce the possibility of overlooking crucial details. Preventative maintenance is less about waiting for things to break and more about anticipating needs based on wear patterns and performance data. This can significantly extend the lifespan of your equipment, ensuring that all tools are ready for use when needed most.

Beyond scheduling, constant monitoring of equipment for signs of wear is essential. This includes looking out for cracked seals, rusted components, and other visible indicators of deterioration. For example, rust around joints and metal fittings in high-humidity environments can lead to failure under stress. Cracked seals might result in leaks, reducing the effectiveness of water delivery through hoses or pipes. Promptly addressing these repairs ensures that each component functions correctly, preserving the integrity of the entire system.

This proactive approach not only maximizes the operational functionality of the equipment but also mitigates liabilities associated with system failures during critical operations. A diligent inspection process can minimize downtime and prevent costly emergency repair measures, enhancing overall workplace safety.

Keeping detailed records of all maintenance activities is not just good practice; it's invaluable in planning and tracking the longevity and reliability of the equipment. These records should include dates of inspections, detailed notes of findings, repairs undertaken, and replaced parts. Proper documentation provides a historical reference that can highlight recurring issues, aiding in future diagnostics and decision-making processes.

Documentation also plays a crucial role in compliance with safety regulations and standards, which may be required by governing bodies overseeing fire safety practices. In addition, meticulously kept records can support warranty claims and inform budget allocations for new acquisitions or replacements. Over time, analysis of this data can reveal trends that may warrant changes in maintenance schedules or introduce new technologies to manage wear and tear more effectively.

Incorporating technology into record-keeping can streamline this process. Digital logs, accessible via cloud-based systems, offer real-time updates and can be easily shared among team members. Such systems improve accuracy and accessibility, allowing firefighting teams to access vital information at the click of a button.

While maintaining the mechanical aspects of firefighting systems is undeniably crucial, understanding the operational context in which these systems are deployed is equally important. Often, the success of a fire response hinges as much on the preparedness and coordination of human resources as on the functionality of the equipment. Comprehensive training programs should accompany maintenance routines, ensuring that all personnel are familiar with procedures and equipment operation.

These programs should incorporate drills and mock scenarios that simulate real-life situations, promoting familiarity with the equipment and its limitations. Engaging in regular training enhances the ability of teams to operate efficiently under pressure, minimizing errors that could compromise both firefighting efforts and personal safety.

Post-Incident Review and Emerging Trends

IN THE EVOLVING LANDSCAPE of firefighting, enhancing fire suppression strategies is critical to ensure safety and operational readiness. This task involves two key dimensions: effective post-incident analyses and staying updated with technological advancements.

Conducting comprehensive post-incident reviews serves as a cornerstone for improving fire suppression strategies. After an incident, gathering data from various sources like reports, debriefs, and video footage is paramount. These diverse data points offer valuable insights into what transpired during the incident. For example, analyzing video footage can reveal tactical decisions that were effective and others that need refinement. By identifying areas needing improvement, firefighting teams can sharpen their tactics for future incidents. It's not just about understanding what went wrong but also recognizing what worked well to reinforce positive actions in future scenarios.

Implementing insights gained from these reviews can significantly enhance training programs and maintenance protocols. Training should evolve to reflect real-world experiences, integrating lessons learned from past incidents. For instance, if a particular tactic repeatedly proves ineffective, revisiting and updating training modules is necessary to incorporate alternative strategies. Adjustments in maintenance protocols may also be required to address any equipment failures or inefficiencies noted during incidents. Integrating such insights ensures that both personnel and equipment are better prepared for future challenges, maintaining a high state of readiness and reliability.

Selecting future equipment also benefits from these post-incident analyses. If a certain piece of equipment underperformed or hindered operations, it provides a strong case for exploring alternatives. Conversely, equipment that excelled can guide procurement decisions, ensuring investments contribute positively to operational efficiency. The National Fire Protection Association shows the importance of technology in reducing house fires through improved sensors and alarms, demonstrating how data-driven decisions can yield tangible safety improvements (Khan et al., 2022).

Keeping abreast of emerging technological trends is another vital aspect of enhancing fire suppression strategies. Recent innovations like smart sensors, robotics, ultra-high-pressure systems, and advanced foam applications present exciting opportunities. Smart sensors, for example, have revolutionized early detection capabilities. They can discern between genuine fires and false alarms, minimizing unnecessary dispatches and ensuring timely responses to actual threats. Robotics offers unique advantages in dangerous or inaccessible environments, allowing firefighters to maintain a safe distance while still addressing hazards effectively.

Ultra-high-pressure systems are transforming traditional fire suppression techniques by delivering more targeted and efficient water application, thus conserving resources while achieving superior results. Advances in foam applications, too, provide new avenues for quenching fires quickly and effectively, particularly in situations where water alone may not suffice. As outlined in the International Association of Fire and Rescue Services report, these advancements are part of a broader emphasis on constructing smart buildings, enhancing both detection and suppression capabilities (Kim et al., 2024).

To effectively integrate these technologies, it's crucial for departments to allocate resources towards researching and adopting the most suitable options. Collaboration with technological developers and engaging in pilot projects can pave the way for smoother transitions and better adoption rates. Furthermore, ongoing training and education about these technologies ensure that firefighters are comfortable and proficient in utilizing new tools.

The integration of informative digital twins (IDT) into emergency management systems exemplifies how technology can enhance situational awareness and improve response times during emergencies. By simulating potential fire scenarios and assessing risks, IDTs help outline safer egress routes and optimal deployment strategies (Kim et al., 2024). Such advancements demonstrate the transformative potential of embracing cutting-edge technology within fire services.

While pursuing these innovations, it's important to recognize that technology is only as effective as its users' ability to deploy it skillfully. Continuous education, practical drills, and simulations facilitate the mastery of new equipment and procedures, ensuring that firefighters can leverage technological advancements to their full potential when facing real-world challenges.

Summary and Reflections

MAINTAINING AND EVALUATING fire suppression systems are vital aspects covered throughout the chapter. By focusing on the meticulous care of hoses and nozzles, regular inspections and cleaning prevent issues from escalating, ensuring these critical components remain functional when needed most. The importance of proper storage methods cannot be overstated, as organized systems reduce wear and tear, prolonging equipment lifespan. Equally significant is the role of training personnel to handle equipment responsibly, fostering a culture of accountability and expertise among firefighting teams. This approach not only safeguards against potential failures but also secures the efficiency and reliability of fire suppression efforts.

In addition to maintenance practices, attention to post-incident reviews and technological advancements is crucial for future readiness. Analyzing past incidents provides insights that refine tactics and highlight necessary adjustments in training and procedures. Embracing new technologies such as smart sensors and robotics can revolutionize firefighting operations, enhancing both safety and effectiveness. However, integrating these innovations requires continuous education and practical experience to ensure firefighters utilize them proficiently in real-life scenarios. By combining traditional strategies with modern advancements, this approach equips firefighting teams to tackle emergencies with confidence and precision, ultimately reinforcing their mission to protect lives and property.

Chapter 10 Questions and Answers

Why is maintaining and evaluating fire suppression systems important for fire departments?

It ensures equipment readiness and operational efficiency, allowing firefighters to be fully prepared during emergencies.

What are the primary components of a fire suppression system that require maintenance?

Hoses, nozzles, foam systems, portable pumps, standpipes, deck guns, and other firefighting appliances and tools.

What should firefighters check during a hose inspection?

They should check for visible signs of wear and tear, such as cracks, weak spots, kinks, abrasions, and issues with seals and nozzle connections.

What cleaning techniques are recommended for fire hoses?

Using mild detergents to wash hoses, ensuring thorough drying to prevent mold, and cleaning after each use to remove debris and residue.

Why is it important to store hoses properly?

Proper coiling, storing in a cool, dry place, and avoiding exposure to direct sunlight, extreme temperatures, and chemicals helps prevent aging and damage.

How should hoses be handled during transport?

They should not be dragged across rough surfaces to avoid abrasions. Protective covers should be used when necessary.

How does regular maintenance of hoses and nozzles benefit fire departments?

It helps prevent malfunctions during emergencies, extends equipment lifespan, and ensures operational readiness.

What is the role of preventative maintenance schedules in fire suppression systems?

They help detect potential issues early, before they escalate into significant problems, ensuring all components remain functional during emergencies.

What should be included in a preventative maintenance routine for firefighting appliances and tools?

Visual checks for damage, verification of pressure levels, and function tests of moving parts.

Why is record-keeping essential for fire suppression system maintenance?

It helps track inspections, repairs, and replacements, allowing for better resource planning and anticipating future maintenance needs.

What common issues should be monitored during appliance and tool inspections?

Cracked seals, rusted components, and signs of wear such as deteriorating moving parts.

How can technology improve the record-keeping process for fire suppression systems?

Digital logs and cloud-based systems offer real-time updates, accessibility, and greater accuracy, improving maintenance tracking.

What role do training programs play in fire suppression system maintenance?

They ensure that personnel are familiar with equipment operations and maintenance routines, enhancing preparedness and minimizing errors during emergencies.

What is the benefit of conducting post-incident reviews?

It helps identify what worked well and what needs improvement, refining tactics and maintenance protocols for future responses.

How can post-incident reviews influence equipment selection for fire departments?

They can highlight underperforming equipment, guiding future purchases toward more reliable and effective tools.

What emerging technologies are transforming fire suppression systems?

Smart sensors, robotics, ultra-high-pressure systems, and advanced foam applications are among the new innovations improving firefighting capabilities.

How do smart sensors enhance fire suppression efforts?

They help detect fires earlier, distinguish between real fires and false alarms, and minimize unnecessary responses.

What is the role of robotics in firefighting?

Robotics allows firefighters to address hazards from a safe distance, especially in dangerous or hard-to-reach environments.

How does the integration of technology like digital twins improve firefighting?

Digital twins simulate potential fire scenarios, helping to improve situational awareness and optimize deployment strategies during emergencies.

What is necessary for successfully integrating new technologies into fire suppression systems?

Ongoing education, practical drills, and training programs ensure that firefighters are proficient in using new tools and technologies in real-world situations.

Reference List

FOSTER, A. (2021, JULY 21). *Fire Extinguishing Methods and Approach*. FCF | Fire, Electrical & Safety Services Australia. https://www.fcfnational.com.au/blog/extinguishing-fires

Staff, F. (2008, August). *Understanding, Anticipating & Avoiding Flashover*. FirefighterNation: Fire Rescue - Firefighting News and Community. https://www.firefighternation.com/training/understanding-anticipating-avoiding-flashover/

Staff, F. E. (2014, June 2). *Understanding and Avoiding a Flashover*. Fire Engineering. https://www.fireengineering.com/firefighting/understanding-and-avoiding-a-flashover/

What Are the 3 Methods for Extinguishing A Fire? (2023, January 24). Hard Hat Training. https://www.hardhattraining.com/what-are-the-3-methods-for-extinguishing-a-fire/?srsltid=AfmBOoqaNRBZddK6LKpH62I3cujy605sCW-XzE4skKSbY6V7yB1oV4gg

Bettinazzi, V. (2023, November 13). *Nozzles 101: Smooth-bore vs. combination/fog nozzle*. FireRescue1. https://www.firerescue1.com/making-the-cut-the-definitive-guide-to-firefighting-tools/nozzles-101-smooth-bore-vs-combination-fog-nozzle

KNOW YOUR NOZZLES. (n.d.). STEADY FIRE TACTICS. https://www.therollsteady.com/pumpschool/knowyournozzles

Staff, F. (2012, June 21). *Nozzle Types, Pros and Cons*. FirefighterNation: Fire Rescue - Firefighting News and Community. https://www.firefighternation.com/lifestyle/nozzle-types-pros-and-cons/

Axelsson, L. (2023, October 29). *GPM vs BTU = FALSE*. The Swedish Firenerd. https://swedishfirenerd.com/blog/gpm-vs-btu-false/

Amara. (2019, November 21). *NFPA 20: Fire pump design.* Consulting - Specifying Engineer. https://www.csemag.com/articles/nfpa-20-fire-pump-design/

Blog Archive» It's the GPM! | Compartment Fire Behavior. (2014). Cfbt-Us.com. https://cfbt-us.com/wordpress/?p=26

Diasana, S. B. (2020, August 18). *How Much Water Pressure Is Required for a Fire Sprinkler System?* QRFS - Thoughts on Fire Blog. https://blog.qrfs.com/358-how-much-water-pressure-is-required-for-fire-sprinkler-systems/

Filkov, A. I., Tihay-Felicelli, V., Masoudvaziri, N., Rush, D., Valencia, A., Wang, Y., Blunck, D. L., Valero, M. M., Kempna, K., Smolka, J., De Beer, J., Campbell-Lochrie, Z., Centeno, F. R., Ibrahim, M. A., Lemmertz, C. K., & Tam, W. C. (2023, October). *A review of thermal exposure and fire spread mechanisms in large outdoor fires and the built environment.* Fire Safety Journal. https://doi.org/10.1016/j.firesaf.2023.103871

Hiong, E., & Yun Ii Go. (2023, September 5). *Large-scale energy storage system: safety and risk assessment.* Sustainable Energy Research. https://doi.org/10.1186/s40807-023-00082-z

Big Considerations for Medium-Diameter Hoseline Purchases - Fire Apparatus: Fire trucks, fire engines, emergency vehicles, and firefighting equipment. (2021). Fireapparatusmagazine.com. https://www.fireapparatusmagazine.com/equipment/firehoses/big-considerations-for-medium-diameter-hoseline-purchases/

Hose, B. (2021, January 5). *Importance of Fire Hose Size.* Puck. https://puck.com/importance-of-fire-hose-size/

Isakson, C. (2024, April 23). *Command & Control Success: Size Matters.* Firehouse.com; Firehouse. https://www.firehouse.com/operations-training/article/53098768/firefighter-keys-to-selecting-the-best-hose-size-to-fight-fires

Selecting Hoselines - Canadian Firefighter Magazine. (2015, March 30). Canadian Firefighter Magazine. https://www.cdnfirefighter.com/choice-of-diameter-can-affect-firefighter-safety-20677/

(2022). Allhandsfire.com. https://www.allhandsfire.com/Categories/Tools-Equipment/Fire-Tools?srsltid=AfmBOoqWYxsi0_eRHHOjGxgNyPDLCAK2JqQd

All Products. (2024). Fire Safety USA. https://firesafetyusa.com/collections/all-products

Firefighting Tools | Axes, Saws, Wrenches & More. (2021). Fire-End. https://fire-end.com/collections/tools?srsltid=AfmBOoo-Y_4NqJ7fti1Cj73PfCCFTrkrsiUZ2o1ZKblb8

Fire Suppression - Monitors & Appliances. (2024). Fire-End. https://fire-end.com/collections/appliances-monitors?srsltid=AfmBOop0Umf_AVcfetjrnym7_CBRRA4

Firefighting Strategies and Tactics - Third Edition. Flashcards. (2024). Quizlet. https://quizlet.com/748348364/firefighting-strategies-and-tactics-third-edition-flash-cards/

Keeping Industrial zones fire safe – Know what to keep in check to avoid fire hazards – Fire Fighting And Rescue equipments. (2024, January 2). Foryourresque.com. https://foryourresque.com/2024/01/02/keeping-industrial-zones-fire-safe-know-what-to-keep-in-check-to-avoid-fire-hazards/

» Fire Control | Compartment Fire Behavior. (2014). Cfbt-Us.com. https://cfbt-us.com/wordpress/?cat=40

Gettemeier, B. (2023, December 5). *Calling the Audible in a Changing Environment.* Firehouse.com; Firehouse. https://www.firehouse.com/operations-training/article/53075913/fire-department-company-officers-preparation-to-change-tactics-on-the-fireground

Riverside Firemen's Retreat – LIFE IN THE LOWER SUSQUEHANNA RIVER WATERSHED. (n.d.). https://www.susquehannawildlife.net/fireattackstrategy-com/

Staff, F. E. (2000, April). *PRECONNECTS: the basics - Fire Engineering: Firefighter Training and Fire Service News, Rescue.* Fire Engineering: Firefighter Training and Fire Service News, Rescue. https://www.fireengineering.com/firefighting-equipment/preconnects-the-basics/

Smith, B. T. (2013, July). *The 4 Phases Of the Stretch: Phase 3.* Firehouse.com; Firehouse. https://www.firehouse.com/operations-training/hoselines-water-appliances/article/10954005/firefighter-training

Tactics for Deck Gun Blitz Attacks. (2024). Fireengineering.com. https://digital.fireengineering.com/fireengineering/october_2024/MobilePagedArticle.action?articleId=2013871

Apparatus Showcase 2022. (2022, November 14). Firehouse.com; Firehouse. https://www.firehouse.com/apparatus/article/21282814/apparatus-showcase-2022[1]

Choosing Hose Loads with Multiple Options for Your Rural Fire Apparatus - Fire Apparatus: Fire trucks, fire engines, emergency vehicles, and firefighting equipment. (2020). Fireapparatusmagazine.com. https://www.fireapparatusmagazine.com/equipment/firehoses/hose-loads-with-multiple-options/

Kirby, M., & Lakamp, T. (2010). *Common hose loads & finishes. Firefighter Nation: Fire Rescue - Firefighting News and Community.* Retrieved from https://www.firefighternation.com/lifestyle/common-hose-loads-finishes-2/

Pierce Manufacturing, Inc. (2024). *Pierce Manufacturing | Custom Fire Trucks, Apparatus & Innovations.* Piercemfg.com. https://doi.org/1067447/1569743143940/P18RD

Staff, F. (2010, February). *Common Hose Loads & Finishes.* Firefighter Nation: Fire Rescue - Firefighting News and Community. https://www.firefighternation.com/lifestyle/common-hose-loads-finishes-2/

---

1.   https://quizlet.com/412888197/chapter-15-handling-hose-video-quiz-2-flash-cards/

Bane, T. (2019, April 1). *StackPath*. Www.firehouse.com. https://www.firehouse.com/operations-training/extinguishers/article/21070393/foam-fundamentals

Colorado Firecamp, Wildland Fire Suppression Tactics Reference Guide. (n.d.). Www.coloradofirecamp.com. https://www.coloradofirecamp.com/suppression-tactics/how-to-attack.html

Editor. (2021, August 30). *The difference between direct and indirect fire line*. Plumas News. https://www.plumasnews.com/the-difference-between-direct-and-indirect-fire-line/

Mark. (2020, November 9). *Firefighting Basics: The Blitz Attack - Fire Engineering: Firefighter Training and Fire Service News, Rescue.* Fire Engineering: Firefighter Training and Fire Service News, Rescue. https://www.fireengineering.com/firefighting/firefighting-basics-the-blitz-attack/

Staff, F. (2008, March). *The As (& Some Bs) of Foam Firefighting*. Firefighter Nation: Fire Rescue - Firefighting News and Community. https://www.firefighternation.com/training/the-as-some-bs-of-foam-firefighting/

fire science chapter 17 fire control Flashcards. (2024). Quizlet. https://quizlet.com/96503235/fire-science-chapter-17-fire-control-flash-cards/

Khan, F., Xu, Z., Sun, J., Khan, F. M., Ahmed, A., & Zhao, Y. (2022, April 26). *Recent Advances in Sensors for Fire Detection*. Sensors. https://doi.org/10.3390/s22093310

Kim, Y.-J., Kim, H., Ha, B., & Kim, W.-T. (2024, August 28). *Advanced fire emergency management based on potential fire risk assessment with informative digital twins*. Automation in Construction; Elsevier BV. https://doi.org/10.1016/j.autcon.2024.105722

Peppard, N. (2024, October 31). *Size-Up Considerations for the First-Due Engine - Structural Firefighting*. Fire Engineering: Firefighter Training and Fire Service News, Rescue.

https://www.fireengineering.com/firefighting/fire-attack/size-up-considerations-for-the-first-due-engine/

Volunteer Fire Brigade Training Module 3 firefighting apparatus, tools and equipment. (2015). SlideShare; Slideshare. https://www.slideshare.net/slideshow/volunteer-fire-brigade-training-module-3-firefighting-apparatus-tools-and-equipment/50343307

jspence@hartindustries.net. (2024, August 22). *Essential Tips for Hose Care, Maintenance, and Storage.* Hart Industries, Inc. https://www.hose.com/blog/33/essential-tips-for-hose-care-maintenance-and-storage

midwesthose. (2024, October 16). *How to Extend the Lifespan of Your Concrete Hose: Maintenance Tips and Best Practices - MidwestHose.com.* MidwestHose.com. https://www.midwesthose.com/how-to-extend-the-lifespan-of-your-concrete-hose-maintenance-tips-and-best-practices/

Also by Eric J. Neal

From the Streets to a Chief
Igniting Excellence: Challenges of a First-Year Fire Chief
Echoes From the Ashes
The Roadmap to Fire Behavior
The Suppression Lesson: Water vs Fire
From the Shadows: SMU's Journey
Hazardous Materials Unveiled
Rescue Operations Strategies for Saving Lives
Firefighter Survival: Tactics for Staying Alive

About the Author

Eric J. Neal is a Fire Chief and Emergency Management Director with a Master's in Public Administration and a Bachelor's in Emergency Management. He has served in departments across Texas and Tennessee, sharing his expertise through mentorship and education. His first year as Fire Chief inspired this book, offering lessons for leaders in fire service.